11160

D0832704

Emergency and Security Lighting Handbook

Also by Michael Neidle

Electrical Installation Technology
Wiring Circuits
Electrical Contracting (Editor)

Emergency and Security Lighting Handbook

Michael Neidle

AMIEE, FIElecIE, IEng(CEI), ASEE(Dipl)

Heinemann Newnes

To beloved Hettie

Heinemann Newnes
An imprint of Heinemann Professional Publishing Ltd
Halley Court, Jordan Hill, Oxford OX2 8EJ

OXFORD LONDON MELBOURNE AUCKLAND
SINGAPORE IBADAN NAIROBI GABORONE KINGSTON

First published 1988

British Library Cataloguing in Publication Data
Neidle, Michael
Emergency and security lighting handbook.
1. Dangerous and hostile environments.
Lighting
I. Title
621.32′2
ISBN 0 434 91436 3

Printed in England by
Richard Clay Ltd, Chichester

Contents

Preface

The Oxford Dictionary defines 'handbook' as a 'short treatise, guidebook'. This definition describes the book's purpose by relating it to emergency and security lighting. Since these are specialized forms of illumination it is felt that a fuller understanding is obtained by including, with explanations, widespread information of the various aspects, emphasis being given to industrial applications. To minimize the burden of crossreference some repetition is included, and to add to the book's usefulness certain worked examples are set out.

In these times of rising crime there is an increasing demand for relevant information, which is met piecemeal by the numerous technical articles and new standards for equipment. Therefore, the aim is to link all this information into a coherent and practical form so as to be of benefit to management, engineers, technicians, students and others who have any degree of responsibility for safety and security. Use of this handbook should thus result in the reduction of accidents and an improvement in the security of premises.

Acknowledgements

The author and publishers are grateful for the help given by the following firms and authorities in the preparation of this book:

British Standards Institution
Chalmers and Mitchell Ltd
Chartered Institution of Building Services Engineers
Chloride Bardic Ltd
City & Guilds of London Institute
Dale Electric Co.
EDL Industries Ltd
Electricity Council
GEC Measurements
Gent Ltd
Institution of Electrical Engineers
JSB Electrical Ltd
Marathon Alcad Ltd
Osram—GEC Ltd
PB Design and Development Ltd
Philips Lighting
Security Ltd
Simplex Ltd
Smiths Industries Environmental Controls
Thorn EMI Lighting Ltd
Tungstone Batteries

1

Lighting fundamentals

The nature of light

A full understanding requires a detailed knowledge of physics. Briefly we can consider light to be a form of radiant energy acting as a waveform emanating from a light source. Light waves are part of the family of electromagnetic waves, travelling at a speed of 300 km/s, at particular wavelengths (*Figure 1.1*). The wavelength is the distance between peaks of the waves of energy.

By means of a simple glass prism it can be seen that 'white light', e.g. daylight, consists of a spectrum of colours, each of which is refracted (bent) by different amounts—red least and blue most (*Figure 1.2*). Ultraviolet and infra-red radiations are, however, invisible.

Ultraviolet radiation is an essential part of fluorescent lighting. Fluorescent tubes are internally coated with a fluorescent powder, possessing

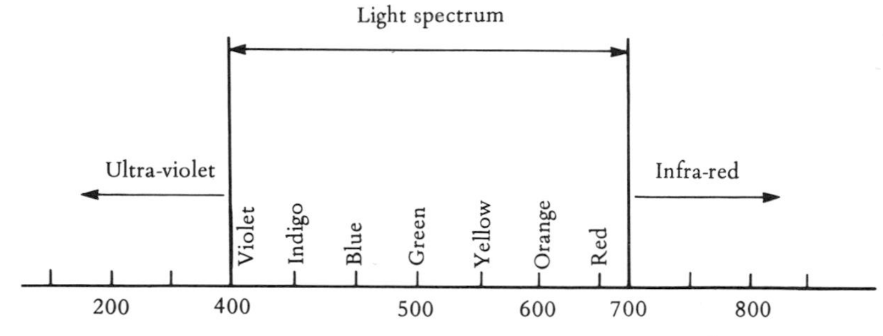

Figure 1.1 *Electromagnetic wavelengths in nanometres (10^{-9} m)*

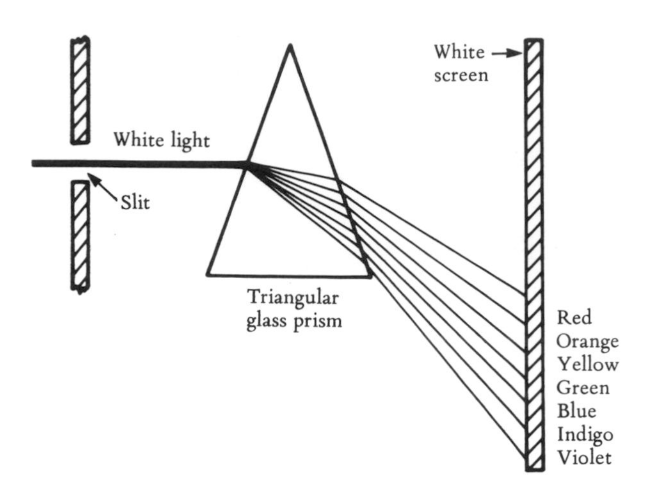

Figure 1.2 *Refraction of light waves*

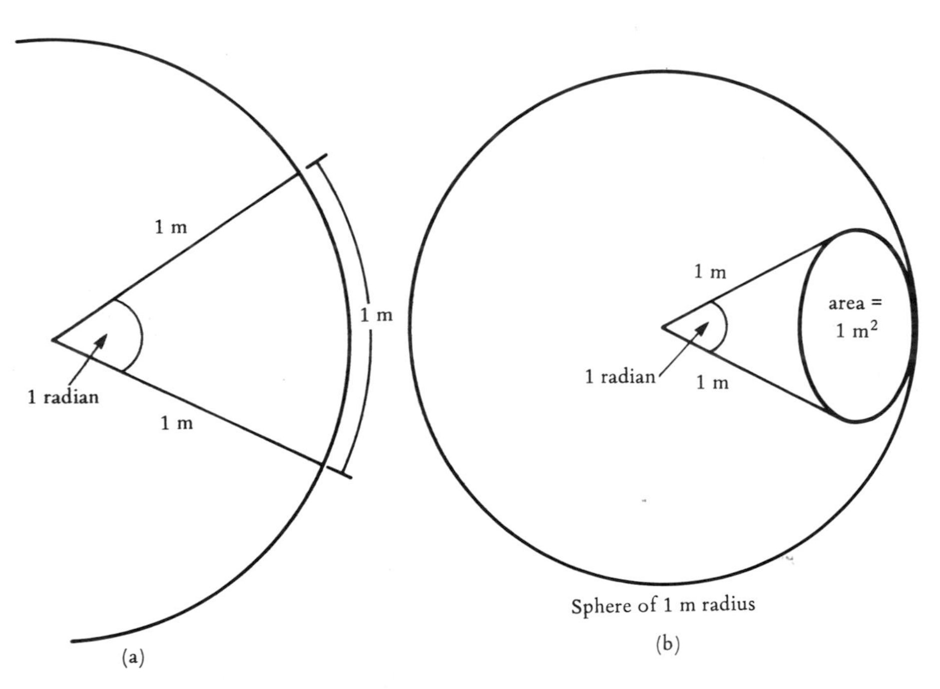

Figure 1.3 *(a) Radian and (b) steradian (solid angle)*

the property of changing the frequency of the ultraviolet waves into radiations within the visible spectrum. Infra-red radiation plays an important part in the design of security lighting.

Lighting units and laws

To ascertain the amount of light necessary for any particular scheme, whether general, emergency or security, an understanding of units and the laws governing these terms is required. The correct relationship between the light source and the surface to be illuminated may then be understood.

1 Luminous intensity (symbol I). The unit is the candela (cd) (pronounced candeela), being the illuminating power of the light source in a given direction.

2 Luminous flux (symbol F). The unit is the lumen (lm) and measures the rate of flow of light emitted from a light source. A lumen is the flux emitted in one steradian (solid angle) from a point source of 1 cd so that the total flux is 4π lm. This can be understood by referring to *Figure 1.3(a)*, where a radian is depicted as the angle subtended by an arc of unit radius of 1 metre (m). A steradian (*Figure 1.3(b)*) is a solid angle enclosing an area equal to the square of the radius.

$$\begin{aligned}\text{Since the surface area of a sphere} &= 4\pi r^2\\ \text{then the number of steradians} &= 4\pi(1)^2\\ &= 4\pi\end{aligned}$$

3 Illuminance (symbol E). The unit is the lux. It shows the density of light falling on a surface and is measured in lumens per square metre (lm/m^2).

4 Inverse square law. Since light travels in straight lines the same light flux from the point source would fall on the surfaces A, B and C (*Figure 1.4*). Surface B is twice as far from the source as surface A, but its area is four times that of A. If the illuminance on A is 1 lux per unit area, then that on surface B is 1/4 lux. Similarly the illuminance at C, which is three times as far from the source as A, is 1/9 lux. This demonstrates the inverse square law as a fixed relationship between source intensity and the

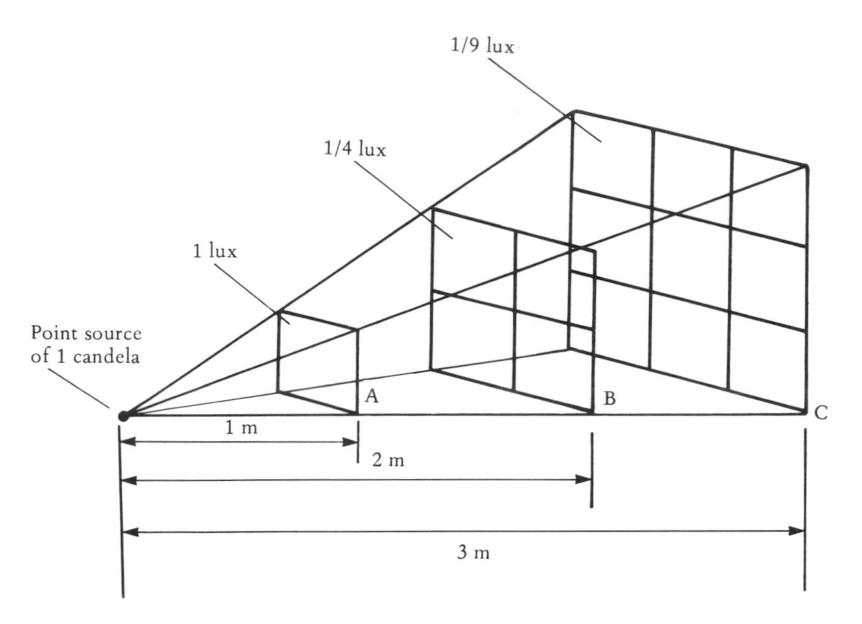

Figure 1.4 *Inverse-square law*

distance between source and illuminated surface. It may be stated in general terms as

$$E = \frac{I}{d^2}$$

where E = illuminance (lux)

I = intensity of source (cd)

d = distance between source and surface to be illuminated (m)

This equation refers to a point source such as given by the common metal filament lamp. With a linear source, e.g. the 1500 mm or 1800 mm fluorescent tube, the inverse law often ceases to apply, in which case the equation is simplified to $E = I/d$ lux.

5 *Cosine law*. We have now to consider the case where the plane to be illuminated is not perpendicular or *normal* to the source. Referring to *Figure 1.5*, A and B are two plane surfaces. B is inclined to A by an angle θ. Comparing areas,

$$\frac{B}{A} = \frac{1}{\cos\theta}$$

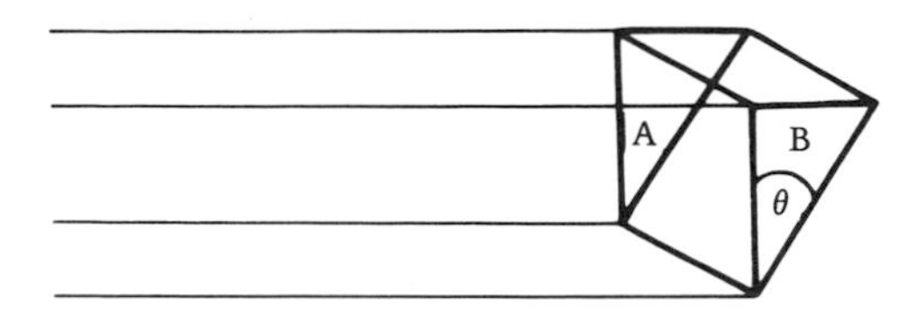

Figure 1.5 *Cosine law*

Assuming light rays are perpendicular to surface A, then the flux falling on B (with A removed) is decreased by $\cos\theta$, and since B presents a larger surface area, the inverse square law can be modified to $E = I \cos\theta/d^2$. The law is of general application. If there is no inclination, $\theta = 0°$ and $\cos 0° = 1$.

6 *Luminaires.* The definition of a luminaire given in the IEE Wiring Regulations is as follows:
'Equipment which distributes, filters or transforms the light from one or more lamps, which includes any parts necessary for supporting, fixing and protecting the lamps, but not the lamps themselves, and, where necessary, circuit auxiliaries together with the means for connecting them to the supply.'

7 *Polar curves.* The intensity distribution of a lamp or lamp with a shade/reflector may be illustrated by a polar curve (*Figure 1.6*) which shows the intensity distribution in a vertical plane. The data for the polar curve is as follows:

Angle in degrees measured through the centre of the lamp cap	0	15	30	45	60	75
Luminous intensity (cd)	280	310	400	480	475	420

8 *Efficacy.* This term indicates lamp efficiency and is measured in lumens per watt (lm/W). In the chapter on lamp types (Chapter 2) we shall see the variety of different efficacies given by various lamp sources. These need to be examined by designers responsible for general, emergency and security lighting.

9 *Coefficient of utilization.* Part of the lumen output of sources is lost in the fittings. Some output may be directed to the walls and ceilings

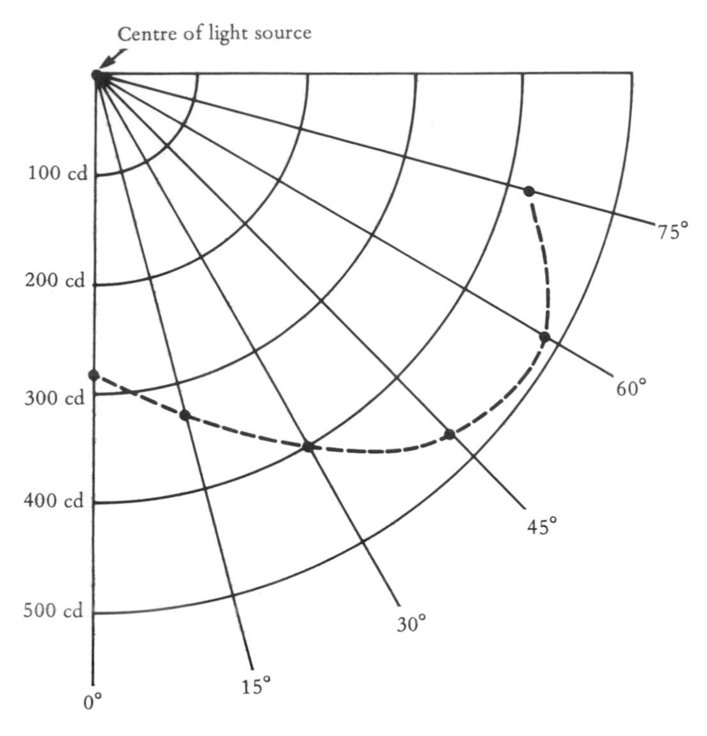

Figure 1.6 *Polar curve of metal filament (GLS) lamp*

where part will be absorbed and part reflected. Thus, only a portion of the light emitted by the lamp reaches the working surface. This proportion is expressed as a number called the utilization factor, which is always less than unity. The lower the utilization factor, the higher the required power of the light source for a given illuminance. The actual determination of the coefficient of utilization requires detailed photometry of the luminaire. The procedure is well described in BS 5225 and the data obtained can be used in calculation procedures given in the CIBSE—Technical Memorandum 5. The UF tables are typically given for a mounting height, range of room sizes and reflectances.

10 Maintenance factor. Dust and dirt on lamp fittings reduce the light output. Illumination will also be impaired by deterioration of reflecting walls (and ceilings). A value of 0.8 is commonly taken but will have to be reduced for dusty and dirty atmospheres as is sometimes found in industrial plants. The maintenance factor has been replaced by the light loss factor, which also takes into account the lamp lumen depreciation and lamp survival rate.

11 Spacing to mounting height ratio (SHR). When above ground level, mounting height and positioning of luminaires is of great importance. For emergency lighting the luminaires must be placed at strategic points and clearly in the line of vision, while security light requires positioning to warn off intruders.

General lighting uses the SHR to ascertain the layout of the luminaires needed for uniform illumination over a given area. This ratio is often employed in conjunction with the lumen method of lighting design. The following equation is used to estimate the total lumens required to illuminate a task area:

$$\text{lumens required} = \frac{\text{illuminance (lux)} \times \text{area of working plane } (\text{m}^2)}{\text{light loss factor (LLF)} \times \text{coefficient of utilization (UF)}}$$

Therefore, the illuminance received from a group of luminaires can be expressed as

$$\frac{\text{total lamp lumens} \times \text{LLF} \times \text{UF}}{\text{area of working plane } (\text{m}^2)}$$

Example 1.1

The illuminance on a surface below a light source varies inversely as the height of the light source above the surface. If the light source is 4 m directly above a working surface, calculate the height to which the light source will have to be moved to double the illuminance on the working surface if the light source is

(a) a filament lamp
(b) a long fluorescent tube.

Solution

(a) By the inverse square law for the point source, the illuminance is

$$E = \frac{I}{d^2} \text{ lux}$$

$$= \frac{I}{4^2}$$

where I = intensity in candelas and d = distance between source and surface to be illuminated in metres.
When E is doubled,

$$2E = \frac{I}{d_2^2} = \frac{2I}{4^2}$$

$$\therefore d_2^2 = \frac{4^2 I}{2I}$$

and $$d_2 = \sqrt{8} = 2.83 \text{ m.}$$

(b) From the distance for a line source,

$$\frac{I}{d_2} = \frac{2I}{4}$$

where d_2 is the new distance,

$$\therefore d_2 = \frac{4I}{2I} = 2 \text{ m.}$$

Example 1.2
A lamp has a luminous intensity of 1200 cd. Calculate the illuminance at

(a) A, which is 6 m directly below the lamp on the ground.
(b) B, 4 m along the ground from A.
The problem is illustrated in the following diagram:

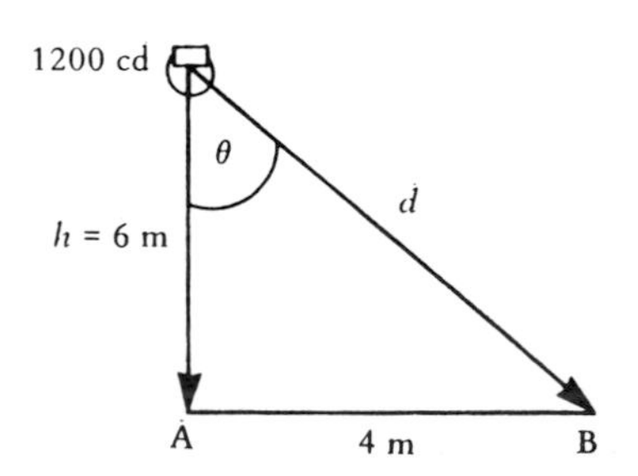

where d is the distance from the lamp to point B and θ is the angle between h and d.

Solution

(a) Illumination at A $= \dfrac{I}{h^2}$

$$= \frac{1200}{6^2} = 33.3 \text{ lux.}$$

(b) From Pythagoras' theorem,

$$d = (6^2 + 4^2)^{1/2} = \sqrt{52} = 7.21 \text{ m}$$

$$\cos\theta = \frac{h}{d} = \frac{6}{7.21}$$

$$E = \frac{I}{d^2} \cos\theta$$

$$= \frac{1200 \times 6}{52 \times 7.21} = 19.2 \text{ lux.}$$

It is assumed that the luminous intensity below the lamp will be uniform in all directions.

The following example indicates an alternative method of applying the cosine law.

Example 1.3

A vertical square wall of side 4.5 m is required to be illuminated. A light fitting fixed to the ground and facing the centre of the wall gives a luminous intensity of 4000 cd in all directions towards the wall at a distance of 6 m from the bottom of the wall. Calculate the illuminance

(a) half-way along the bottom of the wall
(b) half-way along the top of the wall
(c) at one of the top corners

with data shown in *Figure 1.7*.

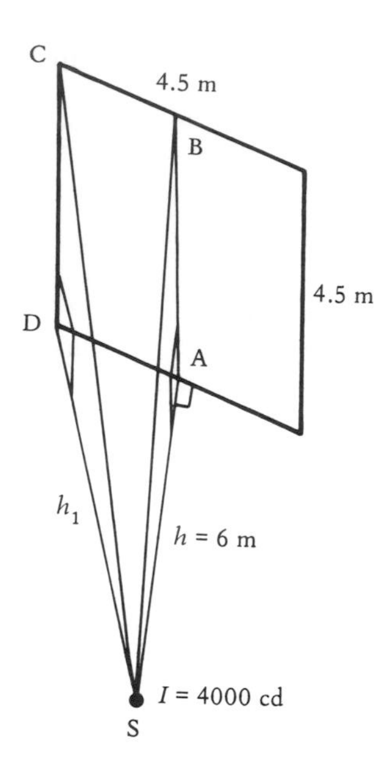

Figure 1.7 *Wall illumination*

Solution

(a) Illumination at the centre of the bottom edge

$$= \frac{I}{h^2}$$

$$= \frac{4000}{6^2} = 111.1 \text{ lux.}$$

(b) From $\triangle$ SAB, $d = 6^2 + 4.5^2 = 7.5$ m. By the cosine law, $E = I/d^2 \cos\theta$, and here $\cos\theta = h/d$

$$\therefore \text{illuminance at } B = \frac{Ih}{d^3}$$

$$= \frac{4000 \times 6}{7.5^3} = 57 \text{ lux.}$$

(c) From SDC, $h_1 = (6^2 + 2.25^2)^{1/2} = 6.4\ \text{m}$

$$d_1 = (6.4^2 \times 4.5^2)^{1/2} = 7.82\ \text{m}$$

$$\therefore \text{illuminance at } C = \frac{Ih}{d_1^3}$$

$$= \frac{4000 \times 6.4}{7.82^3} = 53.53\ \text{lux.}$$

Example 1.4

A store 15 m long, 9 m wide and of 3 m ceiling height is to be illuminated to a level of 300 lux. The utilization and light loss factors are, respectively, 0.9 and 0.8. Make a scale drawing of the plan of the store and set out the required luminaires. Assume a lamp efficacy of 20 lm/W and space-height ratio of unity.

Solution

$$\text{Lumens required} = \frac{\text{illuminance (lux)} \times \text{area (m}^2\text{)}}{\text{LLF} \times \text{CU}}$$

$$= \frac{300 \times 15 \times 9}{0.9 \times 0.8}$$

$$= 56250.$$

SHR = 1, so at a mounting height of 3 m, spacing $= 3 \times 1$
$= 3$ m.

Therefore the number of luminaires required is

$$\frac{15}{3} \times \frac{9}{3} = 15.$$

The lamp units are set out in *Figure 1.8*.
With a lamp efficacy of 20 lm/W,

$$\text{total power} = \frac{56\,250}{20} = 2812.5\ \text{W.}$$

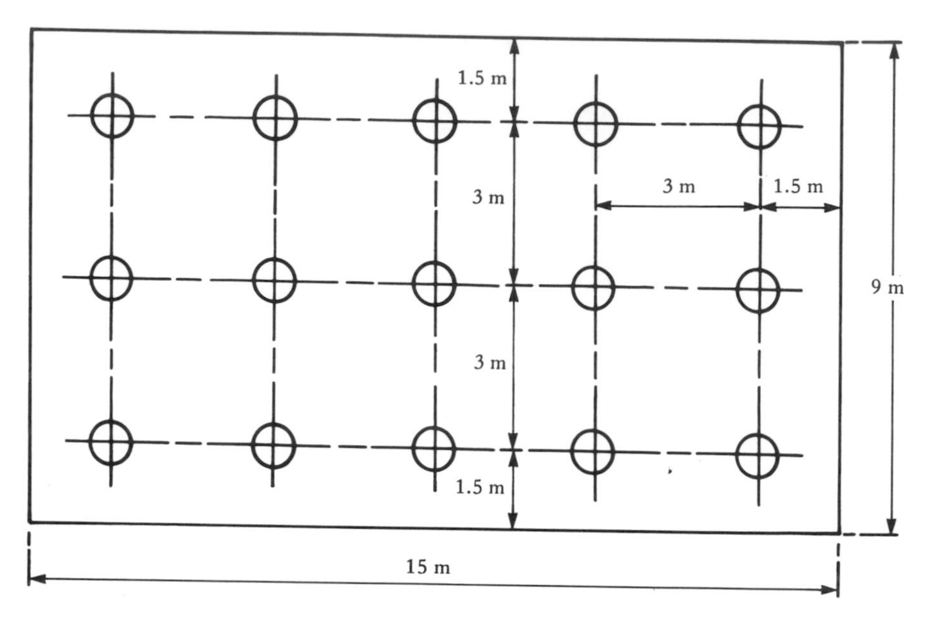

Figure 1.8 *Uniform lighting layout*

As there are 15 lamp units,
power for each lamp

$$= \frac{2812.5}{15}$$

$$= 187.5 \text{ W.}$$

therefore fifteen 200 W lamps are required.

The portable photometer

Sometimes called lightmeter or luxmeter, this device (*Figure 1.9*) often avoids the need for lux calculation. A pointer on the photometer gives a direct reading of the illuminance in lux. The instrument contains a window behind which is housed a photoelectric cell. Light rays falling on the cell produce a flow of electrons. This generated current operates a moving coil-microammeter, the scale of which is calibrated in lux.

Some of the following refinements may be found on these instruments:

1 allowance for light falling on the meter at a wide angle

Figure 1.9 *Portable photometer*

2 the type of source, e.g. whether tungsten or discharge
3 a switch on the instrument to permit high or low readings
4 correction filters for different light colours
5 cosine correction.

Meters should conform to BS 667 1968, 'Portable photoelectric photometers'. To obtain a correct reading they must be placed parallel to the surface to be illuminated.

The GEC minilux 2 photometer (*Figure 1.9*) is particularly suitable for checking the level of emergency lighting installations (BS 5266 Part 1, 1975). At 0.2 lux the dial cannot be read unless it is illuminated with, say, a torch. Photometers may be obtained with up to ten ranges. Typical ranges on a ten-range instrument would be (in lux):

0–0.25
0–1
0–2.5
0–10
0–25
0–100
0–250
0–1000
0–2500
0–10000.

Controls

Control switches must be rated to carry lighting loads with due allowance for reactive currents. Unless specially designed to break inductive circuits, the rating of switches controlling fluorescent, and other discharge, lamps must be reduced by 50 per cent, that is, they must have a rating of twice the current they are required to switch. In circuits with large power factor capacitors there will be high inverse current. Uncorrected lagging circuits will generate high EMF that may flash over the switch contacts. Assuming the control gear power factor is not less than 0.85 lagging, then the demand for fluorescent fittings in volt-amperes is 1.8 times the rated lamp wattage. All control gear must be situated as near to its associated lamp as possible.

Example 1.5

A single-phase supply consists of sixty 65 W fluorescent tubes and thirty 100 W general service lamps. Assuming no diversity, and a supply voltage of 240 V, what size of main switch is required?

Solution

Current allowance for fluorescent tubes $= \dfrac{1.8p}{V}$

$$= \frac{1.8 \times 60 \times 65}{240}$$

$$= 29.25 \text{ A}$$

Current allowance for filament lamps $= \dfrac{P}{V}$

$$= \frac{30 \times 100}{240}$$

$$= 12.5 \text{ A}$$

Total current $= 29.25 + 12.5 = 41.75$ A

The nearest available main switch size is 50 A. This will allow for a number of extensions.

The following switches are often used in simple lighting circuits:

1 the 1-pole, 1-way switch, which is used to break only the phase conductor
2 the 1-pole, 2-way switch (a changeover switch). This is used to control one lamp from two positions or, as illustrated in *Figure 1.10*, to switch between two sets of lights.
3 the series-parallel switch. This was originally designed as a 3-heat switch for the control of cooker plates. It may be used in lighting circuits to give three degrees of illumination, as shown in *Figure 1.11*.

The Lighting Industry Federation offers advice on semiconductor switching. The switching device must conform to BS 800 to avoid the

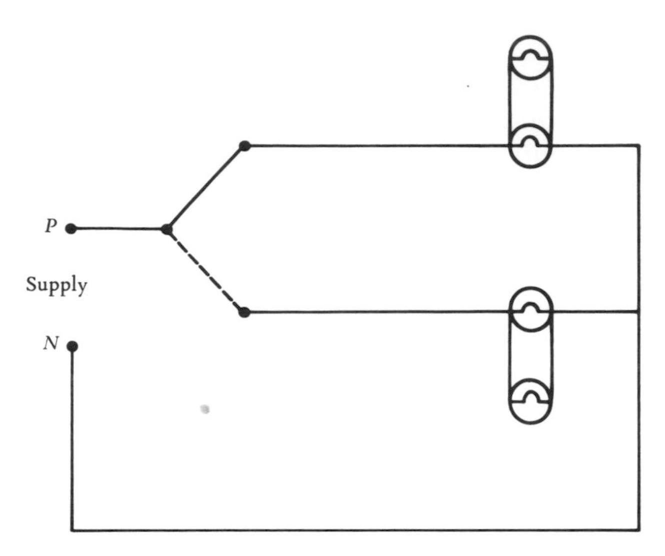

Figure 1.10 *2-way change-over switch*

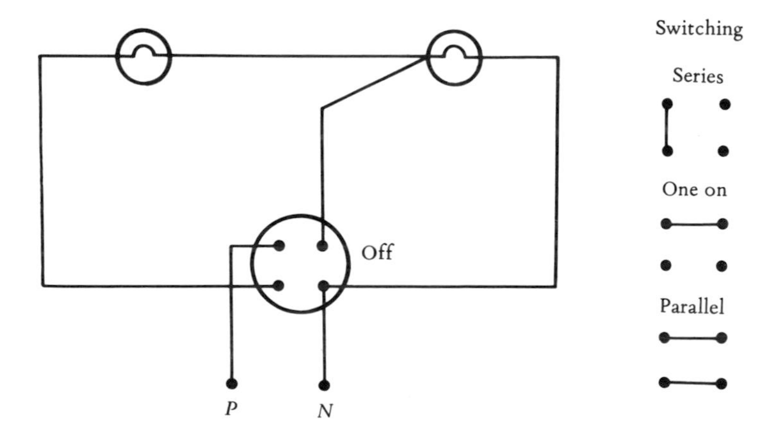

Figure 1.11 *Series-parallel switching*

generation of radio interference. Care must be taken that these semiconductor switches are adequately separated from both luminaire components and wiring to avoid adverse effects.

Switching by remote control does not require any cables between the switch and lamp (*Figure 1.12*). The hand-held control transmitter emits infra-red signals which are received by the ceiling- or wall-mounted master. A press button gives a simple 'on-off' or dimmer action by means

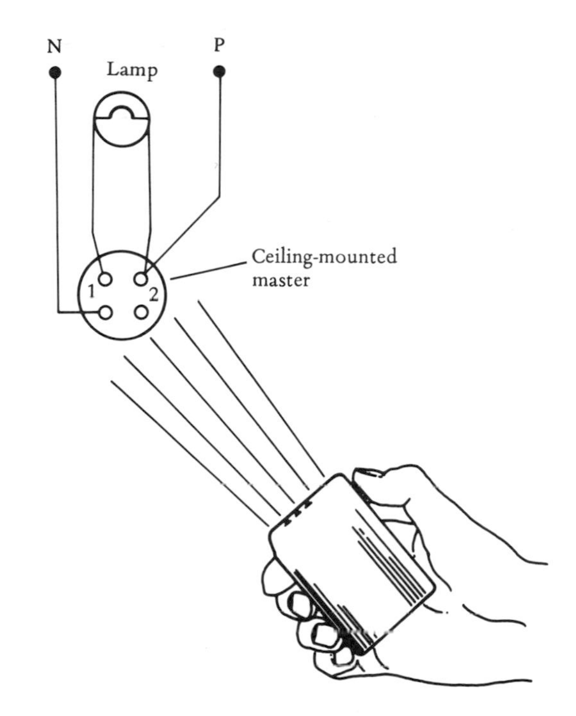

Figure 1.12 *Light control by infra-red rays*

of a single pulse. The infra-red rays will pass through glass with a range up to 10 m. Further wall-mounted extension 'slave' units may be added by connection to terminals 1 and 2.

Safety

All wiring for emergency and security lighting must comply with the current edition (and amendments) of the 'Regulations for Electrical Installations', popularly known as the 'IEE Wiring Regulations'. Part I states fundamental safety requirements which may be summarized as relating to comprehensive aspects of design, planning, inspection and testing. They are applicable to both permanent and temporary installations. These regulations may also be used in association with directives issued by the European Economic Community (EEC). One such directive is known as the 'low-voltage directive of 1973' and covers all electrical equipment, with certain specified exceptions, designed for use with a voltage rating of between 50 V and 1000 V AC or between 75 V and 1500 V DC.

There is no statutory requirement in the UK that an electrical installation has to comply with the IEE Wiring Regulations, but an installation that does comply with these regulations is deemed to satisfy the requirements of regulation 27 of the statutory 'Electricity Supply Regulations 1937'. All installations, however, are subject to regulation 26 of these Electricity Supply Regulations, which place a statutory duty on the supply undertaking 'not to permanently connect a consumer's installation with their electric lines unless they are reasonably satisfied that the connection, if made, would not cause a leakage from the consumer's installation exceeding one ten-thousandth part of the maximum current to be supplied to the said installation'.

As an indication of fire hazards, it is often claimed that many outbreaks of fires in Britain are caused by electrical faults. Owing to possible heat rise, luminaires must be fixed to prevent ignition of materials placed in proximity. Luminaires may be mounted on flammable surfaces, and shades and guards should also be able to withstand heat rise.

It is also necessary for luminaires to comply with British Standards as applicable. Conformity is indicated by the kitemark emblem (*Figure 1.13*) and can be accepted in conjunction with BS 4533 'Electrical Luminaires'. The use of British Standards Institution's (BSI) kitemark is wide-ranging, covering quality control and performance of luminaire component parts in detail.

Industrial wiring and equipment must be in accordance with 'Electricity

Figure 1.13 *BSI kitemark*

(Factory Act) Special Regulations 1908 and 1944'. This may be replaced by the general revision of the 'Health and Safety at Work Act' with the intention of making new electrical regulations which will also impose requirements for hospitals, teaching establishments, hotels, offices, shops, theatres, research and railway establishments.

Equipment with the BSI safety mark (*Figure 1.14*)

Figure 1.14 *BSI safety mark*

satisfies both BS 4533, as mentioned above, and the UK/EEC safety requirements for lighting equipment.

Both the British Standards Institution's kite and safety marks form part of their Safety Mark Licence scheme. Before a manufacturer of lighting fittings is granted such a licence the following considerations must be taken into account:

1 that the design complies with the appropriate British Standard and involves submitting samples for testing by the BSI

2 quality control and continuous testing procedures and technical staff must be such as to ensure that all latter products are in no way inferior to the samples as submitted (BS 5750).

Even after the licence is granted, checks will be made by regular and unannounced visits to the manufacturer's factory by a member of the BSI inspectorate staff. The Consumer Protection Act and the Electrical Equipment (Safety) Regulations are further examples of the increase in UK legislation with the aim of protecting users of electrical equipment.

Lighting in hazardous conditions

The British Standards Institution is also associated with a certification scheme for emergency lighting equipment to ensure that it conforms with Industrial Committee for Emergency Lighting (ICEL) 1001: 1978 (Per-

ICEL
CERTIFICATION SCHEME
NO. 10011
BS REPORT NO. ....

Figure 1.15 *ICEL certification*

formance of Battery Operated Emergency Light Equipment). The certification emblem is shown in *Figure 1.15*. Some 40 of the leading UK manufacturers subscribe to the ICEL scheme, which was formed from an amalgamation of the Lighting Industry Federation and the Association of Manufacturers allied to the Electrical and Electronic Industry (AMA). A typical lighting escape luminaire subscribing to this scheme may be seen in *Figure 1.16*.

Flammable gases which may explode when ignited by an electric spark probably constitute one of the most severe hazards. Where there is a risk of explosion, Regulation 27 of the Statutory Electricity (Factories) Special Regulation, 1908 and 1944, states 'All conductors and apparatus exposed to the weather, wet, corrosion, *flammable surroundings* or *explosive atmosphere* (our italics) . . . shall be so constructed, or protected, and such special precautions shall be taken as may be necessary adequately to prevent danger in view of such exposures or use'.

Certified flameproof electrical apparatus and fittings will prevent the transmission of flame such as will ignite flammable gases (*Figure 1.17* illustrates a flameproof luminaire).

Flameproof gear formerly bore an emblem in the form of a crown outlining the letters FLP. Owing to the increasing complexity of potentially dangerous situations, this simple (at one time all-embracing) emblem has been superseded by a host of letterings which are more informative.

Hazardous areas are categorized into zones:

Zone 0 is an area where an explosive gas is always present, or for prolonged periods.

Zone 1 is an area where the flammable material is processed, handled

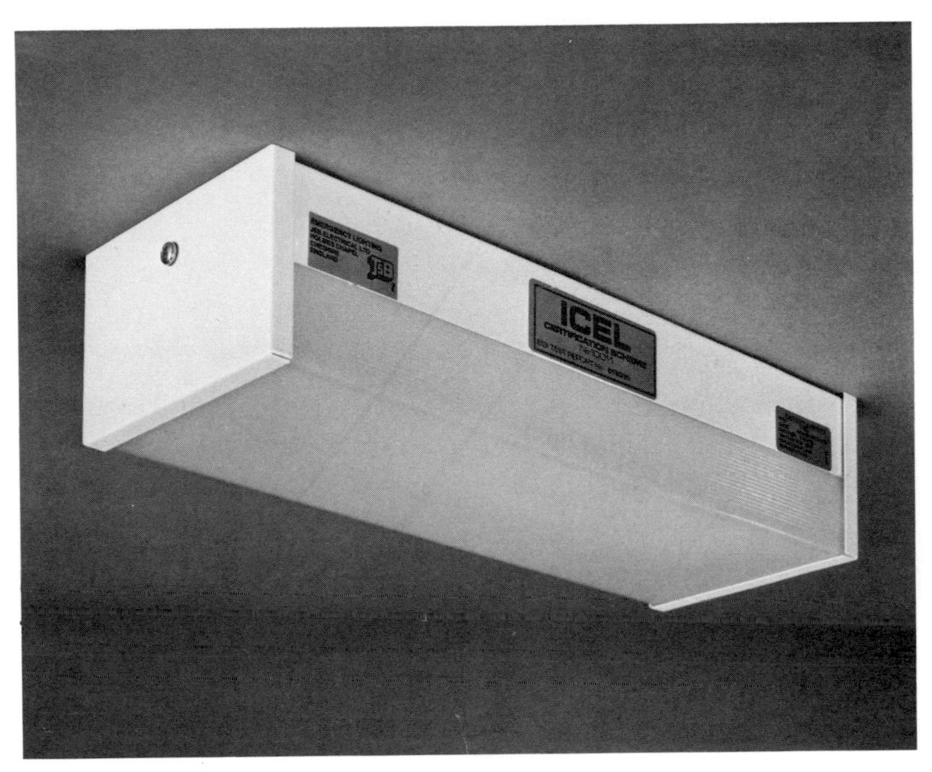

Figure 1.16 *Escape luminaire with ICEL emblem*

and stored, and there is a flammable atmosphere only during normal operation.

Zone 2 is an area where the flammable material, although processed or stored, is controlled so that a flammable atmosphere is only likely to occur during abnormal conditions and only for a short time.

In the United Kingdom, equipment must comply with the requirements of the British Approvals Service for Electrical Equipment in Flammable Atmospheres (BASEEFA). FLP (*Figure 1.18(a)*) is now replaced by Ex. European Economic Community (EEC) countries place Ex inside a hexagon (*Figure 1.18(b)*).

Additions to the required type of protection are:

Ex d flameproof enclosure
Ex l increased safety
Ex ia increased safety (higher category)
Ex ib increased safety (lower category)
Ex o oil immersion

Figure 1.17 *Flameproof luminaire for hazardous area (zone 1)*

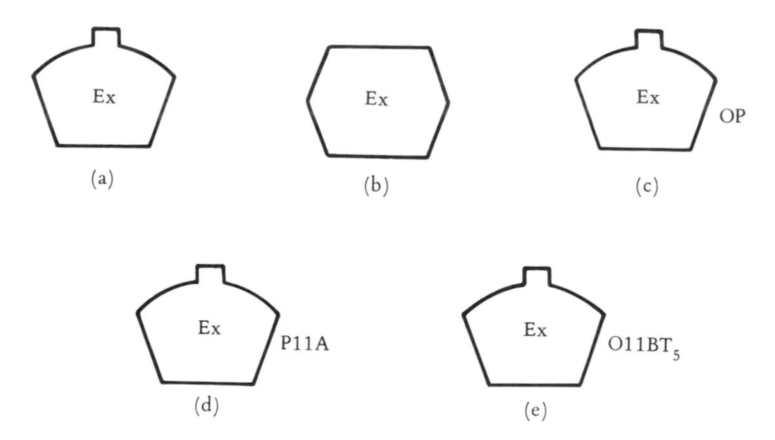

Figure 1.18 *Equipment symbols for hazardous areas*

Ex p pressurized
Ex q sand or quartz filler
Ex s special protection
Ex n other types of protection

Categories ia and ib give *intrinsic safety*, which is defined as 'a circuit in which no spark or any thermal effect produced in the test conditions prescribed in BS 5501 (which include normal operation and specified fault conditions) is capable of causing ignition of a given explosive atmosphere'.

Pressurization as a form of protection against the entry of explosive gases is achieved by introducing air, CO_2 or an inert gas under a slight pressure to the interior of fittings. Protection for zone 0 is met by Ex ia. Protection for zone 1 is given by Ex b, Ex d, Ex l, Ex p or Ex s, while Ex n, Ex o or Ex q is suitable for zone 2.

Where more than one type of protection is necessary, the symbol must appear outside the crown, with the main protection concept given first. *Figure 1.18(c)* illustrates oil and pressurization.

Explosive gases are also classified under group numbers: Group I is gas encountered in coal-mining such as methane (firedamp); group II includes various gases met in industry, such as cellulose vapour, petrol, benzine, amyl acetate etc.; group III is coal and coke gas and ethylene oxide; group IV covers excluded gases, i.e. where there is no flameproof approval such as for acetylene, carbon disulphide and hydrogen, although approval may be granted in individual cases. A full list is given in BS 229. Apparatus marked group II can be further subdivided into IIA (propane), IIB (ethylene) and IIC (hydrogen). As an example of this classification *Figure 1.18(d)* stands for pressurized group IIA gas.

Clearly, surface temperatures are of vital importance. The following is extracted from IEC Publication 79, Electrical Apparatus for Explosive Gas Atmospheres:

classification	maximum surface temperature (°C)
T_1	450
T_2	300
T_3	200
T_4	135
T_5	100
T_6	85

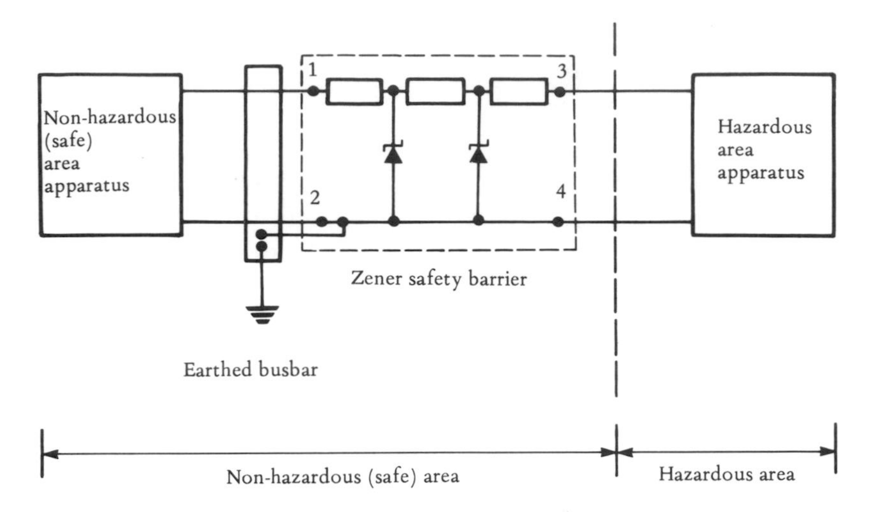

Figure 1.19 *Barrier network*

Figure 1.18(e) may be read as 'oil-immersed group IIB with a maximum surface temperature of 100°C'.

The zener safety barrier circuit (*Figure 1.19*) is designed to limit the energy which can be transferred from a safe area to a hazardous area even under faulty conditions. The zener safety barrier incorporates a resistor of value such that the area is short-circuit-proof. The fuse protection will only become operative either in the event of incorrect polarity connection in the safe area or if an excessive voltage is applied to terminals 1 and 2.

At least two zener diodes are used so that the zener barrier assembly remains safe under component failure conditions. The fuse is further incorporated so that it is possible to determine a maximum design power for the zener diodes. The base terminals 2 and 4 must be securely bonded to earth.

Working practice for equipment in hazardous areas is covered by BS 5345, 'Code of practice for the election, installation and maintenance of electrical apparatus in potentially explosive atmospheres (other than mining applications or explosive processing and manufacture)'.

2

Lamp selection

Since electric lighting has been with us for over a hundred years, it is not surprising that there is a multitude of lamp types. Improvements to existing types and new forms are being continually developed, with the trend towards compactness and increased efficiency.

The emergency and security engineer will look for the lamps that are most suitable for his installation. Factors to be considered are:

1 initial and running costs
2 light output
3 efficacy in terms of lumens/watt
4 maintenance requirements.

General lighting service (GLS) lamps

The ordinary GLS light bulb is one of the most common of all manufactured articles. Many millions are made each year, and as far back as 1952 a leading British lamp manufacturer asserted that their output of glass bulbs was more than enough to throw a girdle of glass, when put side to side, round the earth at the equator! Owing to comparatively low cost and simplicity it is unlikely that this type of filament lamp will ever be completely replaced, despite other light sources with greater efficiency being available.

When current flows through a wire, the energy required to overcome resistance inevitably generates heat. At first it takes the form of long-wavelength heat rays, but with increasing current the wire emits further shorter-wavelength heat by radiation and is heated to incandescence; hence the term 'incandescent' lamp is sometimes used. The filament will now be hot enough to radiate light. While light is produced at the same time it will be accompanied by wasteful heat energy.

Most conductors will either melt or disintegrate at temperatures elevated enough to cause incandescence. Tungsten is chosen for the filament owing to its high melting point—above 3300°C. Yet at about 2100° even tungsten commences to evaporate and the interior of the glass bulb tends to blacken by filament particles flying to the glass walls. Filling the lamp with an inert gas reduces the rate of filament evaporation.

These high temperatures have a profound effect on lampholders and the flexible cord or cable terminal connections. For safety, holders must be of the heat-resisting type and connections must also be able to withstand the heat generated by the lamp. The Ashley lampholder made to BS 5042 Part I T_2 meets higher heat requirements and takes the strain off the terminals.

The weakest link in the vast chain from the power station to the lamp is probably at the bifurcation point above the lamp. In this connection lampholder grips are being employed increasingly to prevent the flexible cord sheath from shrinking back.

A fuse in one or both internal connecting wires is a valuable safety feature, especially in the 'cap-down' position. Fracture of the filament at the end of its life could easily lead to a short-circuit, with the potential danger of plunging several rooms into darkness.

GLS lamp data

The range of the common pear-shaped lamps is from 15 W to 1000 W. *Table 2.1* gives relevant information concerning the average lumen output over the lamp's life. Initially the output would be higher and would decrease to below this average towards the end of the lamp's life.

GLS lamp life

Most manufacturers state a life of 1000 h for the 240 V type, although this period would be shortened by frequent switching on and off. Lamps are very sensitive to supply voltage. When fed by a voltage below 240 V the

Table 2.1 *GLS lumen output*

Power (W)	*Average lumen output* (lm)	*Efficacy* (lm/W)
15	120	8.0
25	215	8.6
40	417	10.4
60	710	11.8
75	900	12.0
100	1330	13.3
150	2140	14.2
200	2880	14.4
300	4550	15.2
500	8200	16.4
750	13100	17.5
1000	18400	18.4

lamp life is increased, and vice versa. Lamp life is also reduced by orienting in any position away from 'cap up', and when the lamp is subject to vibration. Under this latter condition 'rough service' types are recommended. GLS bulbs are sometimes listed as 'double life'—it may be assumed that the extra life is obtained by under-running.

Crompton's claim an average life of 2500 h for their tungsten GLS lamps. More precise effects of voltage changes on standard lamps are:

1 10% voltage increase gives 40% life decrease and 15% increase in light output
2 10% voltage decrease results in 80% life increase and 16% lumen output decrease.

Figure 2.1 illustrates general GLS lamp characteristics. For exact values lamp manufacturers' catalogues should be consulted.

Tungsten halogen (TH) lamp

This is a development of the GLS type. By inserting a trace of halogen, the linear-type lamp (*Figure 2.2*) is of small dimension, e.g. the 300 W version has overall length and diameter, respectively, of 11.8 cm and 1.2 cm and emits an intense white light. Compared with the GLS counter-

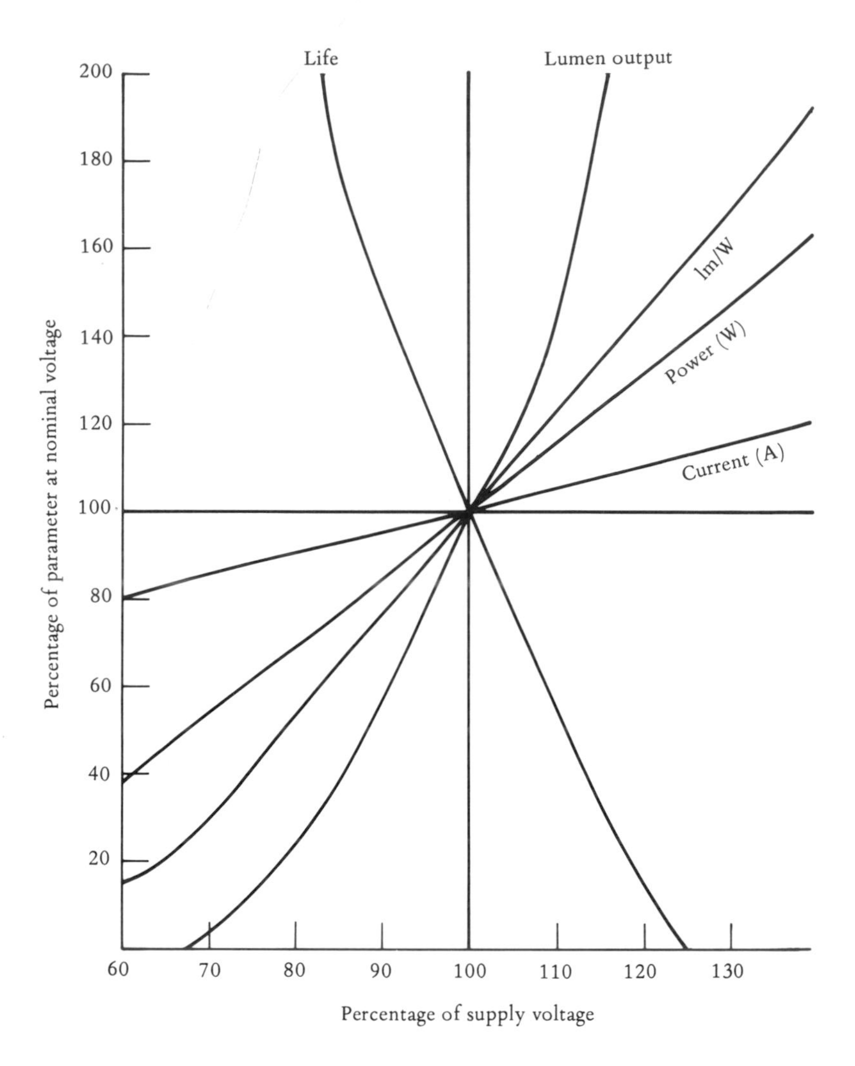

Figure 2.1 *Typical GLS lamp characteristics*

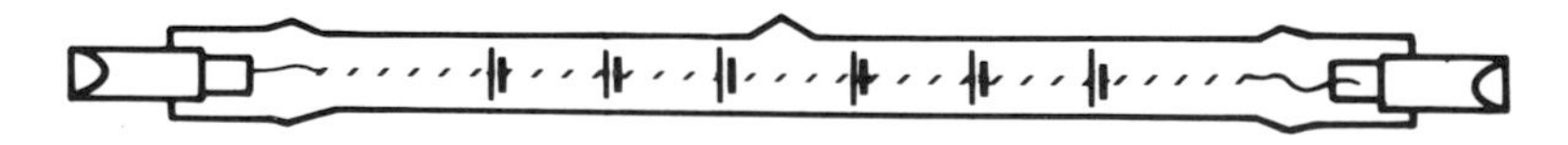

Figure 2.2 *Tungsten halogen lamp*

Table 2.2 *Linear halogen lamps*

Power (W)	*Lumen output* (lm)	
100	1350	lamp life 4000 h universal mounting
150	2100	
200	3100	
250	4000	
300	5000	
500	9500	lamp life 2000 h horizontal mounting
750	15000	
1000	21000	
1500	33000	
2000	44000	

part it can show a 30% increase in light (lumen) output for a given power than for the pure tungsten counterpart, in which particles of the filament are thrown out during its use. By contrast, the halogen lamp follows a regenerative cycle and the evaporated particles are actually restored to the filament, thereby resulting in an increased lamp life (*Table 2.2*) and the prevention of internal lamp blackening.

The table also shows an improved efficacy which ranges from 13.5 lm/W to 22 lm/W. Characteristics of tungsten halogen lamps make them eminently suitable for the floodlighting of large areas as required by security lighting. In use, certain handling precautions are necessary:

1 The range of lamps from 500 W to 2000 W must not be mounted more than plus or minus 4 degrees from the horizontal, otherwise the halogen vapour moves to the lower end and there is rapid bulb blackening at the lower end.

2 The envelope of the quartz glass envelope should not be contaminated by handling with bare hands, so gloves should be used for fitting the lamp into its luminaire. Traces of grease may be removed by methylated spirits or by a similar solvent.

Tubular fluorescent lamps (MCF)

In common with other forms of discharge lighting they operate on a fundamentally different principle from the filament lamps as previously

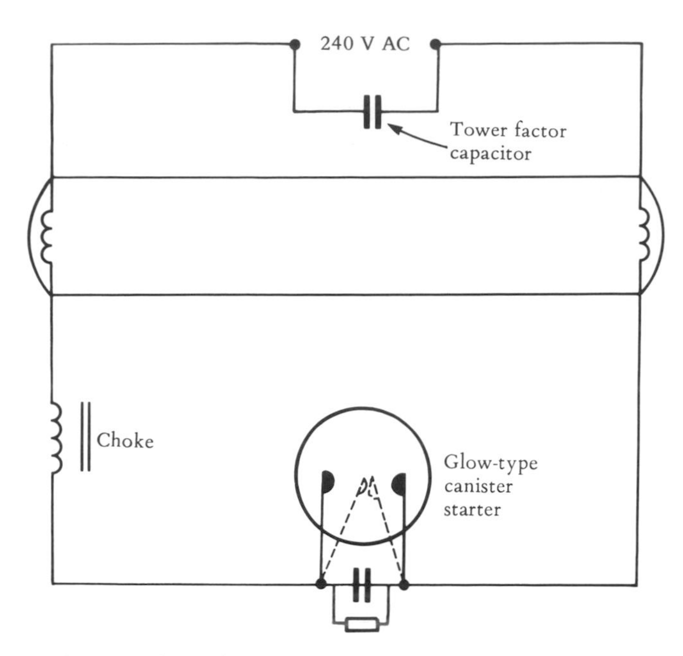

Figure 2.3 *Circuit of standard tubular fluorescent lamp*

described. Here illumination relies on striking an arc between the ends of the glass tube containing a gas or metallic vapour, to which argon may be added to assist starting.

With the popular glow-type starter containing helium gas (*Figure 2.3*) there are three stages for initiating the discharge. The supply current flowing between the starter bimetal strips heats up the helium gas and the cathode coils at the end of the lamp, and at the same time an intense magnetic field is built up in the inductive choke coil. The warming action causes the bimetal strips to make contact, thereby cooling the helium gas since during this period no current flows through the gas. The contacts of the cooled strips now snap apart, thus producing a sudden collapse of the choke's magnetic field. This rapid rate of change induces a voltage which is sufficient to strike an arc across the end of the tube.

The low-pressure vapour discharge is in the form of ultraviolet with some visible radiation, and the tube's internal fluorescent coating transforms the arc into visible light. The actual colour of the light output depends on the chemical employed for the fluorescent material with white or warm white as the most common colour.

Table 2.3 *Tubular fluorescent lamps*

Length	*Power*	*Lumen output (White)*	*Efficacy*
mm (ft)	W	lm	lm/W
600 (2)	20	1000	50
1200 (4)	40	2700	67.5
1500 (5)	80	4600	57.5
1800 (6)	85	5600	66
2400 (8)	125	8800	70

Life—typically 15000 h but dependent on switching frequency. Doubling the switching rate can halve the life

Choking coils, often termed 'ballasts', now perform a secondary function. Fluorescent lamps have the characteristic of reducing their electrical resistance as current flows. Without the choke the current would continually increase to reach short-circuit conditions and probably shatter the lamp. A resistance coil could be fitted in place of the choke but would result in wasteful heat.

Initial flickering may have various causes. Cold weather can make for difficulties in striking the discharge arc between the tube ends. If the snap opening of the bimetal contacts does not coincide with the peak of the AC wave, starting could be delayed. However, modern electronic starters and fluorescent tubes have done much to reduce the starting period, so that the irritating flashes at commencement are now of short duration. Quick starting is also facilitated by using tubes with internal metal strips and earthing of the lamp caps. Instant-start fluorescent fittings are also available which use electronic starters or HF ballast.

To offset the lagging current produced by the choke, a capacitor is connected across the mains solely for power-factor improvement, while a small capacitor across the bimetal strips minimizes radio interference.

Developments

Fluorescent lighting has been available for nearly half a century, so that it is not surprising that although the standard units, as previously described, have served well, it is understandable that lamp manufacturers continually seek further improvements that this form of illumination offers.

While the principle underlying fluorescent lighting has not changed fundamentally, developments follow many directions, namely:

1 slimmer tubes
2 internal coating and gas filling
3 better lighting output
4 quicker starting by solid-state equipment improved power factor (*Figure 2.4*)

These improvements have resulted in giving an efficacy of some 78–80 lm/W with a longer life and 30% reduction in running costs.

The standard 38 mm (1½ in) tube has been retained for certain forms of emergency lighting; otherwise it has been replaced mainly by the T8 26 mm (1 in) type, but also by the 100 W 2400 mm (8 ft) tube. The dimensions of the simple end pins are unaltered. Tube filling is now often by krypton gas, which has replaced argon. The lamp still needs mercury for the vapour discharge. Modern triphosphors as commonly found in TV tubes: the primary colours of red, green and blue are amalgamated to produce white light. The wavelengths of this spectrum mix is 610 nm (red), 540 nm (green) and 450 nm (blue).

Advantage is taken of the tube characteristic in that its efficacy increases with lamp supply frequency and is claimed to be as high as 90 lm/W. A (high) frequency of 28 000 Hz is achieved by a combination of solid-state bridge rectifier, oscillator and ignitor, which also permits running off a 240 V DC supply, which may be necessary for emergency lighting. The arrangement makes for quick starting, even at freezing point, without flicker and does not require a capacitor for near-unity power factor. Further, any stroboscopic effect is eliminated, and there is reduced lumen output deterioration. However, note must be taken of IEE Wiring regulation 613-7 which stipulates that electronic devices must be disconnected when carrying out a 'megger' test for insulation resistance.

Typical sizes of krypton-filled lamps:

26 mm-diameter	18 W	600 mm (2 ft)
	36 W	1200 mm (4 ft)
	58 W	1500 mm (5 ft)
	70 W	1800 mm (6 ft)
38 mm-diameter	100 W	2400 mm (8 ft)

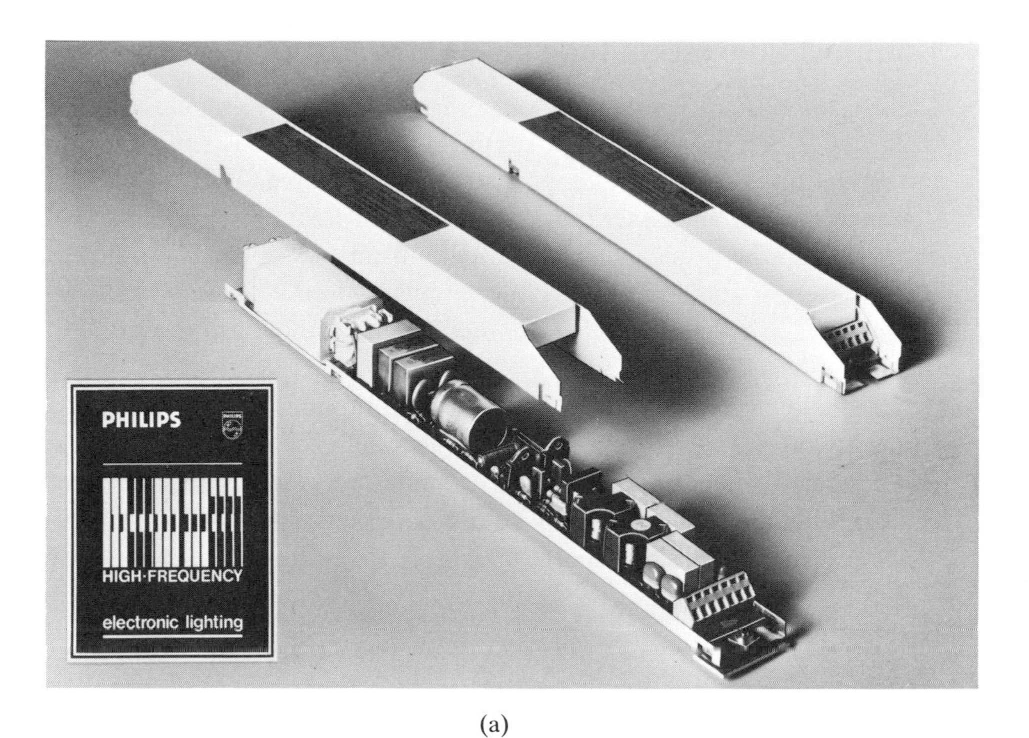

(a)

Lamp 1

Ph

AC mains

N

Electronic filter (preregulated)

Energy storage capacitor

Inverter

Output stage

Lamp 2

(b)

Figure 2.4 *(a) High-frequency electronic fluorescent tube starting gear and (b) HF electronic ballast (THORN)*

Compact fluorescent tube (SL, PL or 2D)

Clearly fluorescent lighting is more efficient than that obtained by the ordinary GLS tungsten filament lamps. However, in many situations the latter has the advantage of a near point light source. With respect to dimension, this new generation of fluorescent lighting forms a radical change from the previous linear fluorescent tubes. An early attempt to reduce overall dimensions was the circular 40–60 W type which was housed in a bowl-shaped glass luminaire, and although these are still available the appeal is limited.

The first compact lamp was obtained by forming the tube into a near square with overall dimensions of not more than 205 mm, including the control gear. Where suitably housed it could form a direct replacement of the GLS bulb. By taking advantage of fluorescent developments as previously described, this new source takes about a quarter of the energy (see *Table 2.4*) with five times the lamp life. The values are typical and may vary with different manufacturers.

Table 2.4 *Power comparison of compact and GLS lamps*

Compact lamp (W)	*GLS lamp (W)*
9	40
13	60
18	75
23	100

Immediate possible uses are for table or bench fitments (*Figure 2.5*), ceiling flush and surface downlighters and floor standards. The long life and small physical dimensions make them eminently suitable for security lighting. When externally fitted they should conform to IP 43 (the digit 4 signifies electrical protection against contact by tools or wires greater than 1 mm thick, while the number 3 indicates protection against rain).

Alternatively there is a wide range of doubled-over U-shaped single-ended compact tubes (*Figure 2.6*). As an example, one rated at 23 W has a very slim tube diameter of 13 mm and an overall length of 236 mm. Two tubes may be connected in series; end caps can be BC or ES to facilitate replacement.

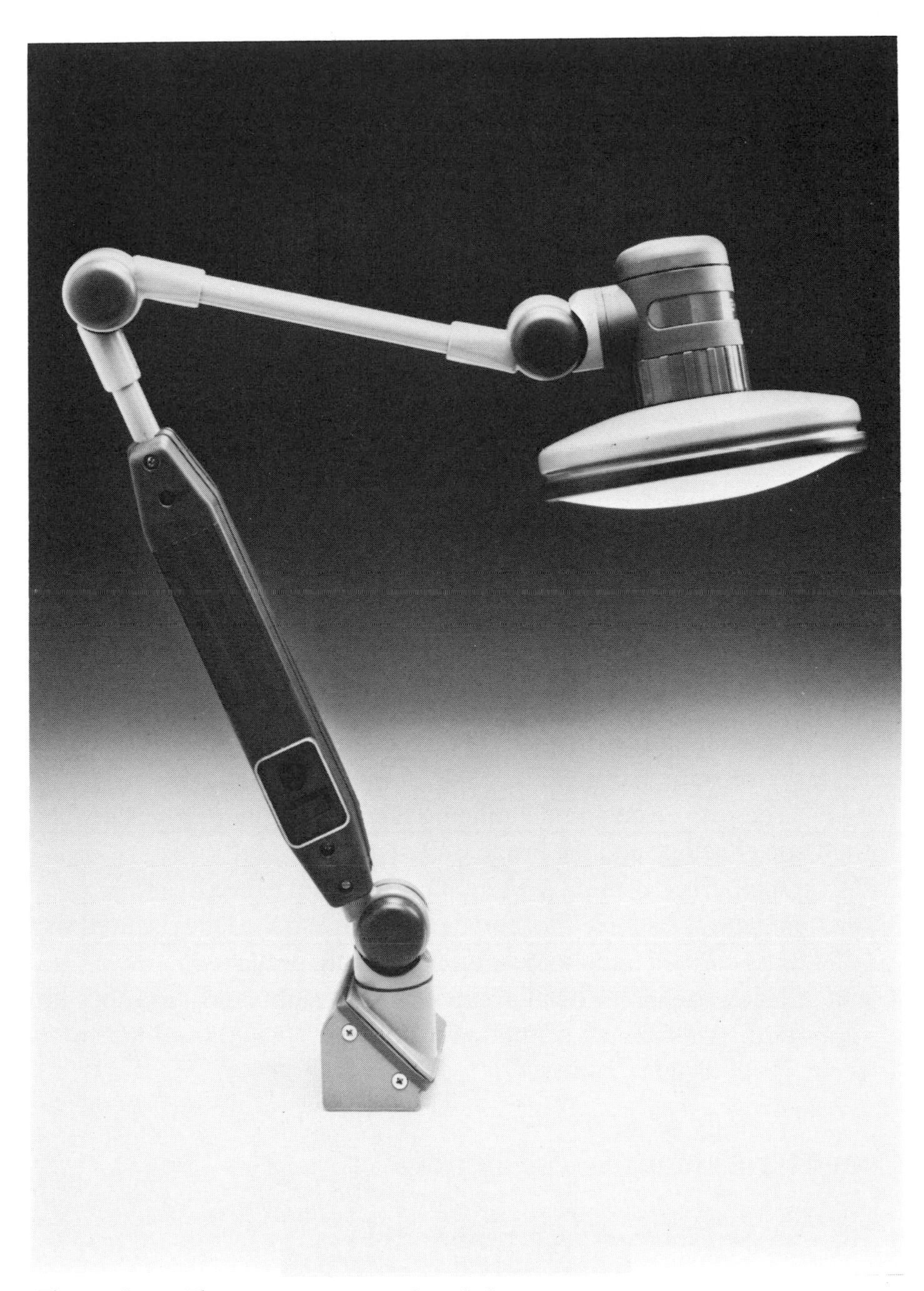

Figure 2.5 *Fluorescent compact bench fitment*

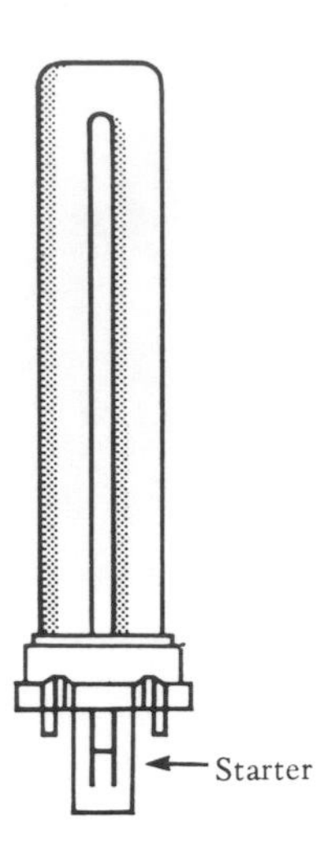

Figure 2.6 *Single-ended compact fluorescent tube*

It is important to note that these compact tubes have a power factor P/VI of approximately 0.8 lagging. This effect may be offset by a group circuit connection of 15.20 μF capacitance per ten lamps.

By virtue of the fact that the base of a compact lamp contains control gear it is heavier and larger than its GLS counterpart. Therefore care is necessary to ensure that:

1 when used with an enclosed luminaire there is adequate space between the lamp envelope and the inside wall of the luminaire
2 the air temperature inside the luminaire does not exceed 75°C
3 the luminaire housing is of adequate strength and the lamp is fitted to a solid brass earthed lampholder; ES holders are preferred
4 where lamps are mounted in a 'cap up' or horizontal position they are supported. An example of the latter position is when used for emergency EXIT signs.

Other forms of discharge lighting

1 High-pressure mercury vapour lamp (MBF)

The fluorescent lamps as previously described are filled with low-pressure gas. Here, as shown in *Figure 2.7*, the discharge takes place in an inner

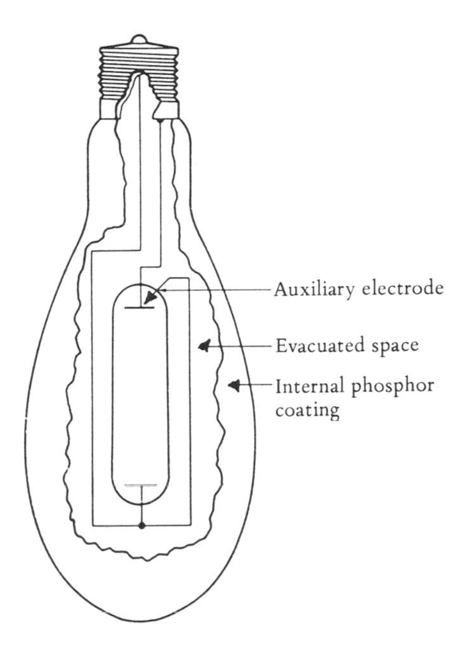

Figure 2.7 *Mercury vapour lamp*

tube of pure silica enclosed in an outer evacuated glass bulb. The inner tube contains some mercury and a small amount of argon to assist starting. In common with other discharge lamps, the electrodes are rich in electron-emitting materials to facilitate electron release. To initiate the discharge, an auxiliary electrode is positioned close to the main electrode. This secondary electrode is connected to the lamp terminal through a high-value resistor to limit the current. On switching on to the supply, the normal mains current is not sufficient, by itself, to initiate the discharge between the main electrodes, and yet it can start over the very short distance between the main and auxiliary electrodes. At this stage, the discharge is in the argon gas. Discharge current passing through the high resistor causes a potential difference, i.e. a voltage, to develop between the starting and main electrodes. This allows the argon discharge to spread until it takes place between the main electrodes, which also has the effect of warming the inner tube and vaporizing the mercury. Soon the gas content becomes mainly mercury vapour and the argon effect is reduced. Discharge after 4–5 min finally occurs in the mercury vapour and the auxiliary electrode takes no further part in the operation of the lamp. The strong bluish-colour output makes the mercury vapour lamp unsuitable for interior illumination, unless modified by an internal fluorescent coating. Applications are in such areas as flood and security lighting. The lamps, which contain internal reflectors or are ovoid in shape, may take a full 5 minutes to reach full brilliance if the supply is cut off, even if the supply is immediately restored. The correct type of lamp must be selected for vertical (V), horizontal (H) or universal (U) mounting. It is emphasized that with an internal phosphor coating the lamp produces a near-white light. Ratings are from 50 to 2000 W. Life expectancy is 8000 to 10 000 h at an efficacy of 35–50 lm/W.

2 Sodium lamp

In one type the low-pressure SOX sodium lamp (*Figure 2.8*) has an inner U-shaped tube made of special glass which is resistant to sodium vapour, since ordinary quartz and hard glass are violently attacked by sodium. A double glass container and high-reactance leak transformer are designed for ease in starting, and there is a vacuum between the two tubes for the purpose of heat conservation.

No starter switch is required as the transformer produces a sufficiently high voltage. The inner tube contains neon gas at low pressure in addition to the sodium and heat produced by an initial neon discharge. At this stage a red glow is emitted due to the neon. Because of the heated discharge, the sodium begins to vaporize, causing the colour of the discharge to change from red to yellow. It takes about 8 to 10 minutes for the full light to be reached, when there is a monochromatic yellow emission, so that objects appear as yellow, black or grey hue, with a long life of 8000–20 000 h.

Down to a temperature of −18°C its starting performance will not be affected and has the advantage of an efficacy of some 190 lm/W, this being greater than that for any other form of artificial electrical illumination.

The high-pressure sodium SON lamp (*Figure 2.9*) is manufactured with an internal silicon-coated bulb, containing the ceramic arc tube which, with the addition of high pressure, broadens the colour spectrum to produce a golden white light. A solid-state ignitor ensures faster starting, to give a full run-up time of 5 minutes. Efficacy is about 110 lm/W, with a life of 20 000 h.

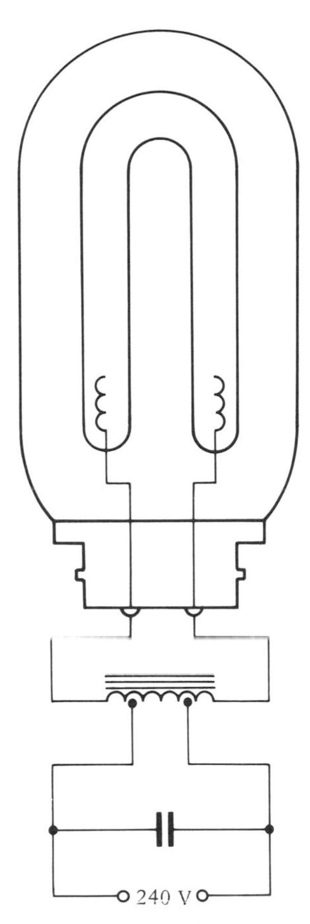

Figure 2.8 *Low-pressure (SOX) sodium lamp*

(Note that all values of life, starting times and efficacies are general since the performance of lamps varies by many factors such as size, colour, rating and the products of different manufacturers.)

3 Metal halide lamp

This is another member of the discharge family of lamps. With or without an internal fluorescent coating it generates excellent colour-rendering white light. The discharge occurs in a high-pressure mercury atmosphere with a metal halide. Full brightness takes place in 2–5 minutes and the efficacy varies from 60 to 80 lm/W. The life expectancy is 5000–10 000 h.

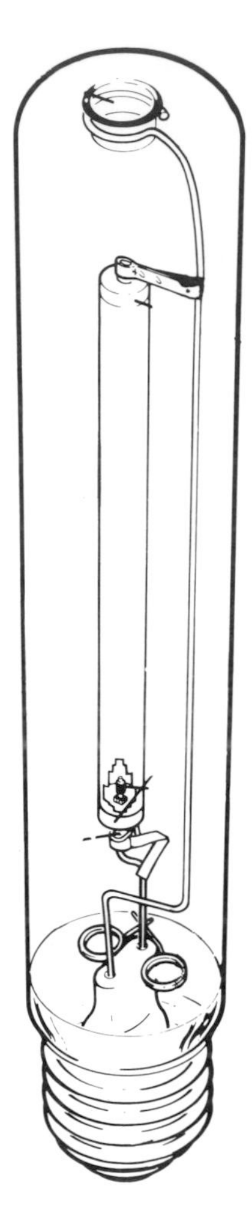

Figure 2.9 *High-pressure (SON) sodium lamp (Osram)*

It gives a very good white light for colour representation. Average values of efficacy (see note above) for various light sources are shown in *Table 2.5*.

Glare and luminaires

Glare may be defined as excessive brightness in the field of vision. Direct light falling on the viewer's eyes makes it difficult to see an object or surface. There may also be the possible effect of high reflection, which is

Table 2.5

Light source	*Efficacy (lm/W)*
GLS	12
tungsten halogen	19
tubular fluorescent	60–90
compact fluorescent	40
low-pressure sodium	190
high-pressure sodium	110
metal halide	70
high-pressure mercury vapour	50

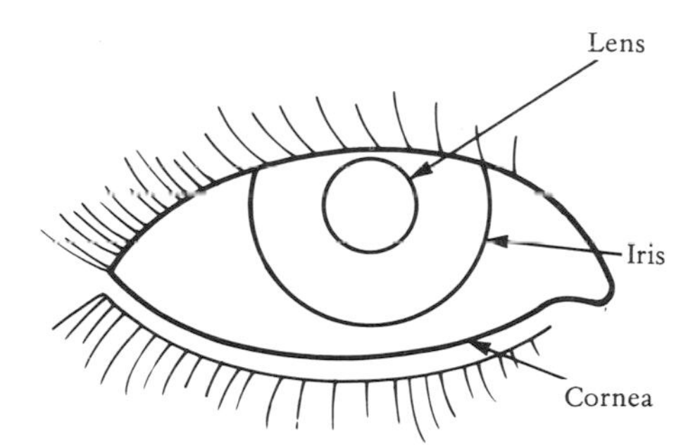

Figure 2.10 *Eye (simplified front view)*

an additional source of glare. For general lighting schemes we require the light to fall on the object or surface—not on the viewer's eyes.

To understand why glare inhibits vision we need to know a little about the action of the human eye (*Figure 2.10*). Light enters the eye through the cornea, the amount being controlled by the iris, acting as a shutter. As we have seen from Chapter 1, light is a form of radiant energy, which passes through the lens to the sensitive nerve layer, the retina, at the back of the eye. It is then conveyed by the optic nerve to the brain, causing the sensation of light. Looking directly at a bright source produces an intense impression on the retina. To avoid damage to this sensitive portion of the eye the iris automatically contracts, thereby reducing the intensity of the image received. Thus, closing of the iris cuts down the amount of lighting received and may be a distinct advantage for security.

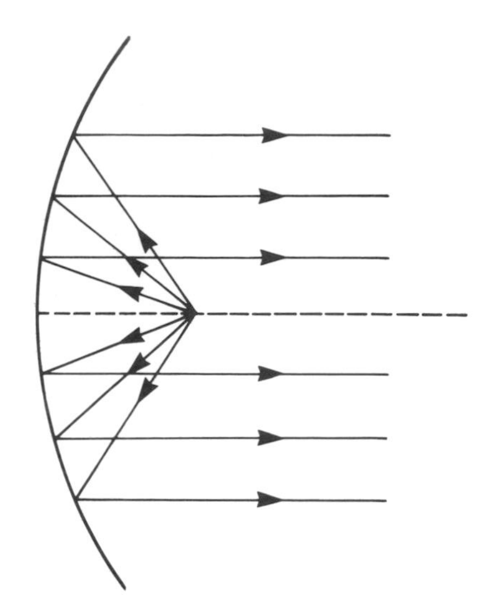

Figure 2.11 *Parallel light beam from parabolic reflector*

Thus, wrong positioning and inappropriately intense lighting actually makes seeing more difficult and also has the deleterious effect of producing eye fatigue. For values of discomfort glare imposed by various situations, based on certain indices, see Technical Memorandum No. 10, issued by the Chartered Institution of Building Services Engineers (CIBSE).

Bare lamps within the line of vision produce glare. It is therefore desirable to spread the light by means of a light fitting. Modern luminaires possess an aesthetic quality, in addition to which they must be so designed and constructed so as not to be affected adversely by the heat of the lamp. For outside use, such as required for security floodlighting of an area, they must also be 'vandal-proof'.

Parallel beams of light are formed by a spherical shape (*Figure 2.11*), but 'batwing' distribution gives a wide spread. Protective methods for luminaires are set out in BS 4533, from which *Table 2.6* has been extracted and is reproduced by permission of BSI. (Complete copies can be obtained from them at Linford Wood, Milton Keynes, MK13 6LE.)

The use of Class 0 luminaires is not permitted in the United Kingdom. While the aim is to avoid glare under normal operating conditions, lighting schemes may be deliberately designed to encourage this phenomenon so as temporarily to blind and scare off would-be intruders.

Table 2.6 *Luminaire protection*

Class	*Description*	*Symbol used to mark luminaires*
0	A luminaire in which protection against electric shock relies upon basic insulation; this implies that there are no means for the connection of accessible conductive parts, if any, to the protective conductor in the fixed wiring of the installation, reliance in the event of a failure of the basic insulation being placed on the environment.	no symbol
I	A luminaire in which protection against electric shock does not rely on basic insulation only, but which includes an additional safety precaution in such a way that means are provided for the connection of accessible conductive parts to the protective (earthing) conductor in the fixed wiring of the installation in such a way that the accessible conductive parts cannot become live in the event of a failure of the basic insulation.	no symbol
II	A luminaire in which protection against electric shock does not rely on basic insulation only, but in which additional safety precautions, such as double insulation or reinforced insulation, are provided, there being no provision for protective earthing or reliance upon installation conditions.	⧈
III	A luminaire in which protection against electric shock relies upon supply at safety extra low voltage (SELV) or in which voltages higher than SELV are not generated. The SELV is defined as a voltage which does not exceed 50 VAC, RMS, between conductors or between any conductor and earth in a circuit which is isolated from the supply mains by such means as a safety isolating transformer or convertor with separate windings.	

Classification of luminaires according to the type of protection provided against electric shock (from BS 4533); reproduced by permission.

3

Power supplies

General principles

Normal lighting is almost universally taken from the standard 240 V AC 50 Hz mains. However, this may not be continuous as cuts and blackouts can occur due to faults in the system, so that emergency and security lighting require alternative electrical supplies in the event of a mains failure. It may be a matter of life or death, e.g. in hospital operations, to ensure that this secondary supply is readily available and switched in automatically without any delay.

Electrical plant selection depends on:

1 the period of time that the alternative supply must be made available
2 reliability
3 defined overload capacity
4 magnitude of load, which for single phase can be calculated from the equation

$$\text{power} = \frac{\text{voltage} \times \text{current} \times \text{power factor}}{1000} \text{ kilowatts}$$

$$\text{or } P = \frac{VI\cos\phi}{1000} \text{kW (n.b. the '1000' changes watts to kilowatts)}$$

The efficiency of the system, i.e. $\frac{\text{output}}{\text{input}}$, must also be taken into account.

Example 3.1

There is a lighting load of 4.5 kW fed from a 120 V single-phase generator supply. If the overall efficiency of the supply is 80%, what is the minimum size of PVC-sheathed cable to carry this load for a run of 10 m? Assume a power factor of 0.85.

Solution

$$P = \frac{VI\cos\phi}{1000}\text{kW}$$

By transposition $$I = \frac{1000\,P}{V\cos\phi}\text{amperes}$$

$$= \frac{1000\times 4.5}{120\times 0.85} = 44\text{ A}$$

Now efficiency $$= \frac{\text{output}}{\text{input}}$$

Working in amperes, $$\frac{80}{100} = \frac{44}{\text{input}}$$

$$\therefore \text{input} = \frac{100\times 44}{80} = 55\text{ A}$$

From IEE 'Regulations for Electrical Installations', Table 9D2, Col. 6,

10 mm^2 cable carries 63 A

Now maximum permitted voltage drop = 2.5% × supply voltage

$$= \frac{2.5\times 120}{100} = 3\text{ V}$$

Actual voltage drop $= \dfrac{\text{mV/A/m}\times I\times l}{1000}$ where l = length of run in metres

$$= \frac{4.4\times 55\times 10}{1000}$$ (mV/A/m obtained from volt drop table, col 3)

$$= 2.42\text{ V}$$

$\therefore$ selected cable size = 10 mm^2.

In practice the type of protection and whether the cable is immersed in lagging must be taken into account (see author's *Electrical Installation Technology*). The two forms of standby or secondary plant are generators and secondary cells (batteries). The standby source must be so arranged as to have no deleterious effect on the normal lighting service. For standby purposes it is necessary to ascertain the characteristics of the source.

Standby generators

The production of electricity by generators is based on the fundamental discovery by the English scientist Michael Faraday. In 1831 he was able to observe that whenever an electrical wire conductor 'cuts' or is crossed by a magnetic field, electricity is generated in the conductor. The principle is illustrated by the up and down movement of a metal rod conductor causing a reading on a sensitive galvanometer (*Figure 3.1*). This action was then developed into an actual generator (*Figure 3.2*). Loops of insulated wire, termed an armature, mounted on a shaft are made to rotate north and south poles of permanent magnets producing an alternating current (AC). For direct current (DC) a commutator is necessary. The electric current so generated is transferred to an external circuit through brass slip rings or a commutator and carbon brushes.

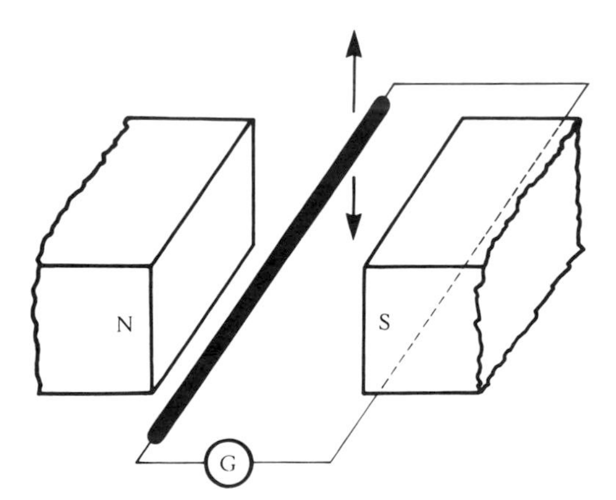

Figure 3.1 *Generation in conductor by up or down movement cutting magnetic field*

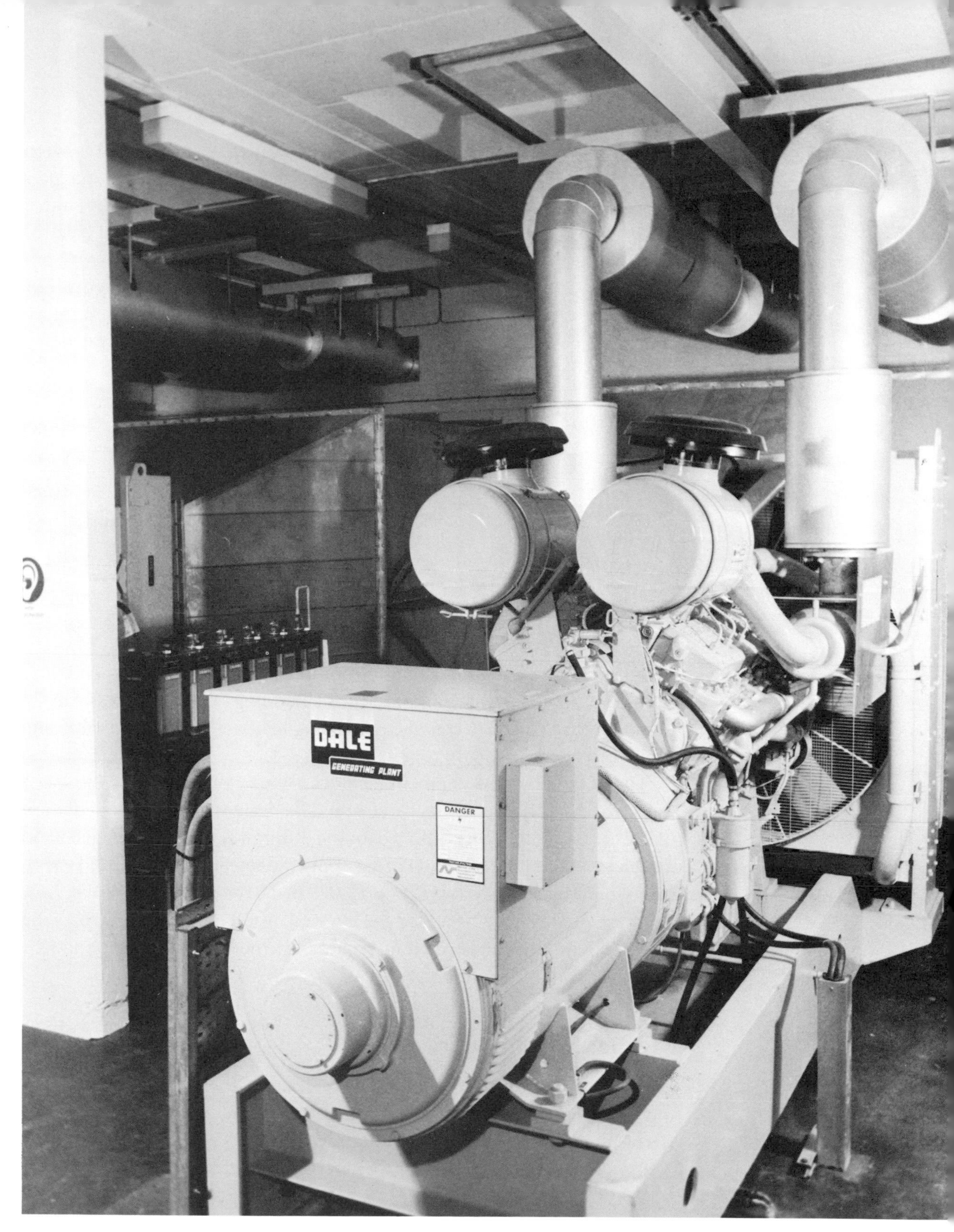

Figure 3.2 *Industrial generator and equipment*

In practice, form generators have many armature windings. It is often found more feasible to reverse the action by rotating pivoted magnets between fixed armature coils. Here the magnets may not be of the simple bar or horse-shoe type and this excitation could be in the form of electro-magnets.

The block diagram in *Figure 3.3* shows the essential arrangements. For maximum efficiency and service, practical forms are, however, more complex. The engine, in smaller sizes, is similar to a lorry motor being fuelled by diesel oil.

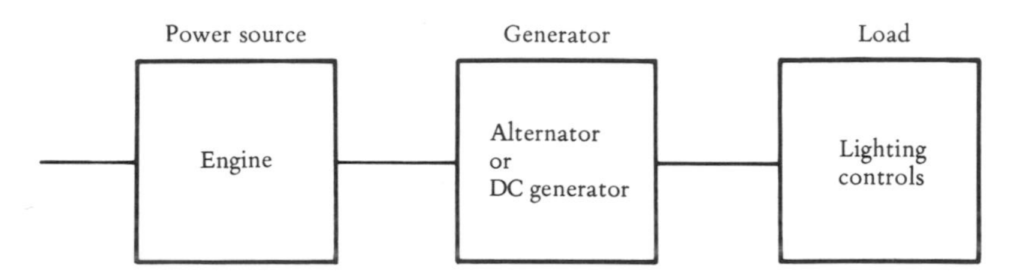

Figure 3.3 *Essentials of generator set*

Spare oil for 6 h running should be made available. Units could be mounted on the same vibration-proof bedding. Gas engines form an alternative to the diesel engine possessing lightweight clean running and may even be mounted on top of a building.

Part of the engine equipment includes a governor to regulate the speed (and, for AC, the frequency) and would be designed to give a regulation of $\pm 2\%$. By regulation is also meant the difference between voltage in no load and full load. Thus for 120 V supply the maximum permitted variation would be 117.6 V to 122.4 V (2.0% of $120 = 2.0/100 \times 120 = 2.4$). To these figures must be added the internal installation voltage drop. The AC generator (often termed alternator) of the modern type would be brushless and self-excited.

Mechanical enclosures

Early generators were of the open type and provided no electrical or mechanical protection. Owing to the obvious dangers they are no longer permitted. Now four main types are recognized (*Table 3.1*).

Table 3.1 *Generator enclosures*

Screen-protected	Covers consist of perforated sheet metal which may easily be removed for inspection purposes. The perforations permit cool air to pass through the windings and other parts of the machine but do not allow the entry of dust or dirt. This protection is recommended for use in clean, dry situations.
Drip-proof	To guard against the effect of dripping water, the perforations are replaced by louvres or cowls which may be backed by a wire mesh to prevent small materials being drawn in by the action of the generator fan.
Pipe-ventilated	Where operating in very dusty or dirty conditions the generator interior would soon become clogged, causing overheating. With the pipe-ventilated enclosure the machine is sealed and ventilation is provided by a pipe system bringing clean air from an external clean atmosphere which may be outside the building.
Totally enclosed	Since the dissipation of heat is only possible through the casing, this type of protection must be given a higher rating for the same output power than one which allows free air circulation.

These divisions are not hard and fast. A recommendation is one which is screen-protected, drip-proof with fan ventilation. Special types would need to be fitted in hazardous situations.

Each generator must have starter control, overload trip and circuit-breaker. The panel, which may be mounted on the generator or adjacent to it, as a minimum, requires instruments showing voltage, current, power and kVA.

When placing an order, care must be taken that spares are readily available, since faults may require replacement of specific parts. Clearly a high degree of reliability will ensure smooth running. Members of the British Generating Set Manufacturers readily give advice ensuring that their equipment follows sound commercial and technical standards.

Maintenance of oil level is vital as running dry will cause destruction of the prime mover engine. Therefore it is essential to fit an oil alarm which automatically stops the engine and gives a positive indication should the magnitude of the oil level decrease to too low a level. In winter the oil should be heated so that rapid run up to full load supply is achieved.

Generator emergency lighting must be able to reach the required illumination level of a minimum of 0.2 lux within 5 s; this may be extended to 15 s at the discretion of the enforcing officer. In practice the level should be above 0.2 lux as allowance must be made for luminaire soiling, voltage drop and ageing of the lamps.

Storage batteries

These are made up of individual cells, silent in running and have no moving parts. It is therefore not surprising that they are in demand where an alternative supply is required to maintain emergency and security lighting. Even with generators, batteries are required for starting purposes. They are often termed secondary cells or accumulators. Here the electrical energy is produced by chemical action. As we shall see there is a variety of chemical materials—giving a wide choice—employed to obtain, in a practical form, the desired voltage. This is not surprising as there has been continuous development for more than a century.

In the primary cell, as used in electric torches, electricity is also generated by chemical action. However the cell has to be discarded when the active materials are used up. This contrasts with the storage battery (consisting of many secondary cells), where the chemical constituents can be reactivated by passing a direct current in opposite direction to the discharge.

While there are other types, batteries fall into two main classes; lead-acid and alkaline. The voltage given out by each cell is small, approximately 2 V for the lead-acid and 1.2 V for the alkaline. For this reason batteries are normally formed by connecting the cells in series. They give the maximum output current when the external connected resistance is equal to the total internal resistance of the cells, often requiring a series-parallel arrangement.

Battery calculations

Calculation of the maximum current may be obtained by:

$$n = \sqrt{\frac{NR}{r}}$$

where n = numbers of cells of each row in series, N = total number of cells, R = external resistance in ohms and r = internal resistance of each cell.

Example 3.2

A battery consists of 24 cells each with an internal resistance of 4 Ω and an electromotive force (EMF) of 1 V. If the external resistance is 6 Ω, show how the cells are connected to produce maximum current. Calculate also the current flowing and the potential difference (PD).

Solution

For maximum current, the number of cells in each row in series is

$$n = \left(\frac{NR}{r}\right)^{1/2}$$

$$n = \left(\frac{24 \times 6}{4}\right)^{1/2}$$

$$n = 6$$

There are thus four parallel groups, each group consisting of six cells in series (*Figure 3.4*)

The overall battery voltage is equal to six cells in series = 6 V

Total internal resistance $= \dfrac{\text{resistance of one row}}{\text{no. of rows in parallel}}$

$= \dfrac{6 \times 4}{4} \qquad = 6\,\Omega$

Total resistance $= 6 + 6 \qquad = 12\,\Omega$

Current flowing $I = \dfrac{6\text{ V}}{12\,\Omega} \qquad = 0.5\text{ A}$

Terminal voltage $= E - (I \times \text{total resistance})$

$= 6 - 0.5 \times 6 \qquad = 3\text{ V}$

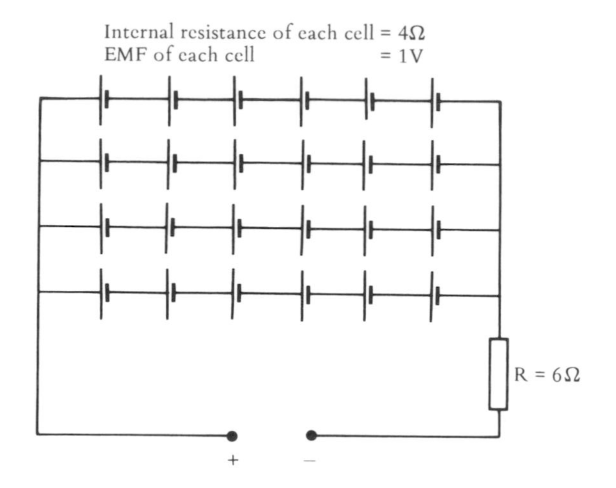

Figure 3.4 *Series-parallel arrangement for maximum current*

Example 3.3

1 State Kirchhoff's Laws

2 A battery of cells with a total EMF of 40 V and a total internal resistance of 2 Ω is connected in parallel with a second battery with an EMF of 44 V and an internal resistance of 4 Ω. A load resistance of 6 Ω is connected across the ends of the parallel circuit. Calculate:

(a) the current in each battery
(b) the current in the load resistor.
Draw a line diagram showing the current directions.

Solution

Current law. When currents meet at a junction, the sum of the currents entering the junction is equal to the sum leaving. In *Figure 3.5(a)*, $I = I_1 + I_2$

$$\therefore \text{ by transposition } I_1 = I - I_2$$
$$I_2 = I - I_1$$
$$O = I_1 + I_2 - I$$

Voltage law. In any closed circuit, the sum of the potential differences

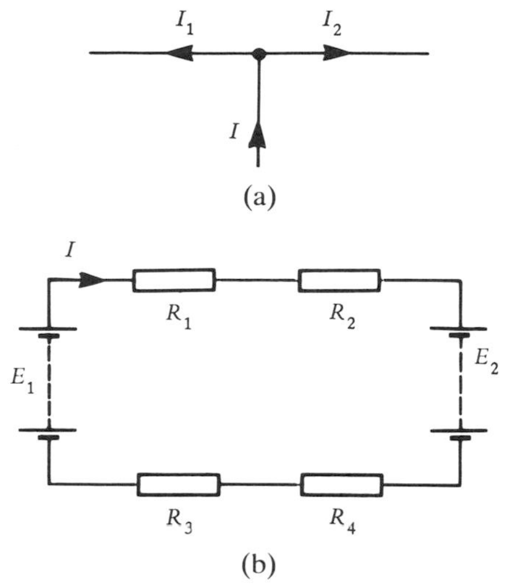

Figure 3.5 *Kirchhoff's (a) current and (b) voltage laws*

(i.e. the products of the currents and the resistances) is equal to the sum of the EMFs (*Figure 3.5(b)*)

$$\therefore E_1 - E_2 = IR_1 + IR_2 + IR_3 + IR_4$$
$$= I(R_1 + R_2 + R_3 + R_4)$$

The convention adopted is to assume that the currents flow round the closed circuit (sometimes called a mesh) in a clockwise direction. This will have to be reversed if the calculation shows a negative sign for the current.

2(a) Inserting assumed currents I_1 and I_2 in *Figure 3.6*, then by the voltage law

mesh ABED $40 - 44 = I_1 \times 2 - I_2 \times 4$

$$\therefore -4 = 2I_1 - 4I_2 \qquad 3.1$$

mesh BCFE $44 = (I_1 + I_2)6 \times I_2 \times 4$

$$\therefore 44 = 6I_1 + 10I_2 \qquad 3.2$$

Equation (1) $\times 3 - 12 = 6I_1 - 12I_2$ 3.3

(2) $\therefore$ (3) $56 = 22I_2$

$$\therefore I_2 = \frac{56}{22} = 2.55 \text{ A}$$

Inverting in (1), $-4 = 2I_1 - 4 \times 2.55$

$\therefore I_1 = 3.1$ A

2(b) Current through resistor R

$= 3.1 + 2.55 = 5.65$ A

The current directions are as shown in *Figure 3.6*.

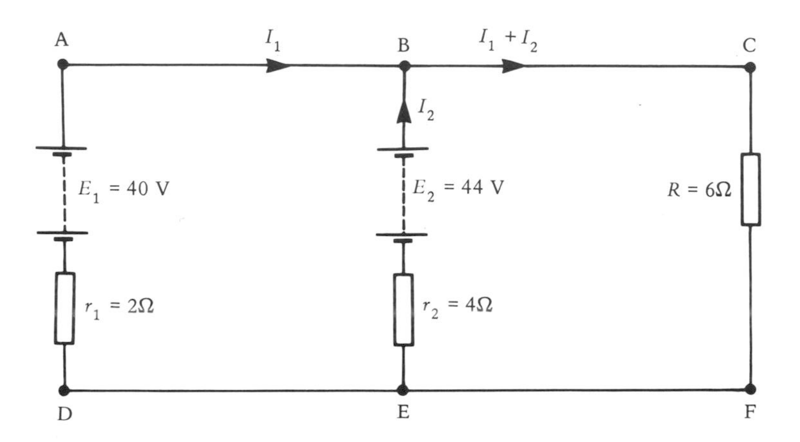

Figure 3.6 *Kirchhoff's law applied to battery calculations*

Lead-acid cell

The lead-acid cell is probably the most common type of secondary cell. It is based on a simple principle: if two lead plates are immersed in a solution of sulphuric acid (called electrolyte) and are charged by connecting to a DC source, a chemical action takes place and the cell becomes a source of DC electrical energy. In modern design many plates are interleaved, each plate consisting of a lead alloy framework packed with active material consisting of a lead paste held in grids or tubes.

During discharge the active material of both plates tends to turn to lead sulphate and the specific gravity of the electrolyte is lowered. Here the specific gravity refers to the weight of a volume of the electrolyte compared to the weight of an equal volume of water and is measured by a hydrometer (*Figure 3.7*).

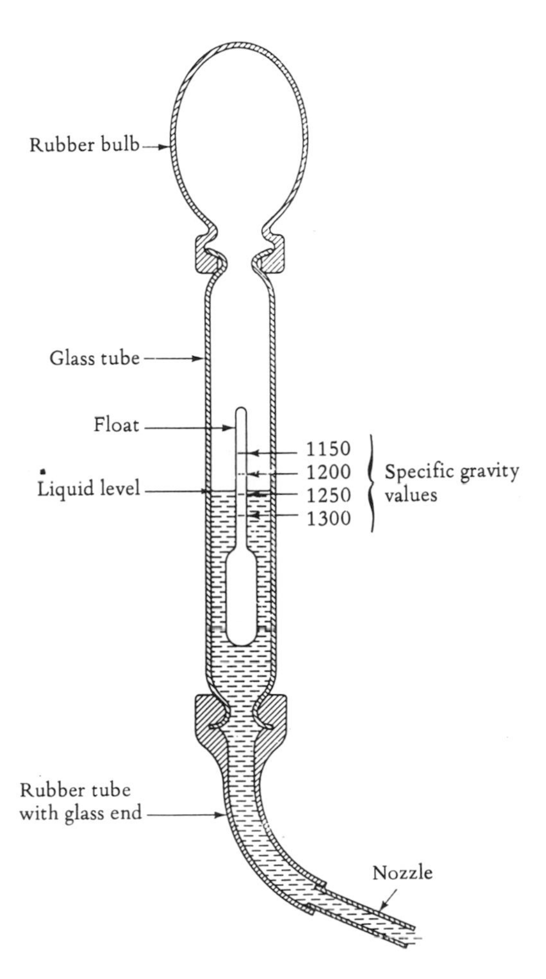

Figure 3.7 *Hydrometer*

The characteristics of the lead-acid cell may be summarized as approximately:

	EMF (V)	*Specific gravity*
charged cell	2.6	1.28
uncharged cell	1.9	1.15

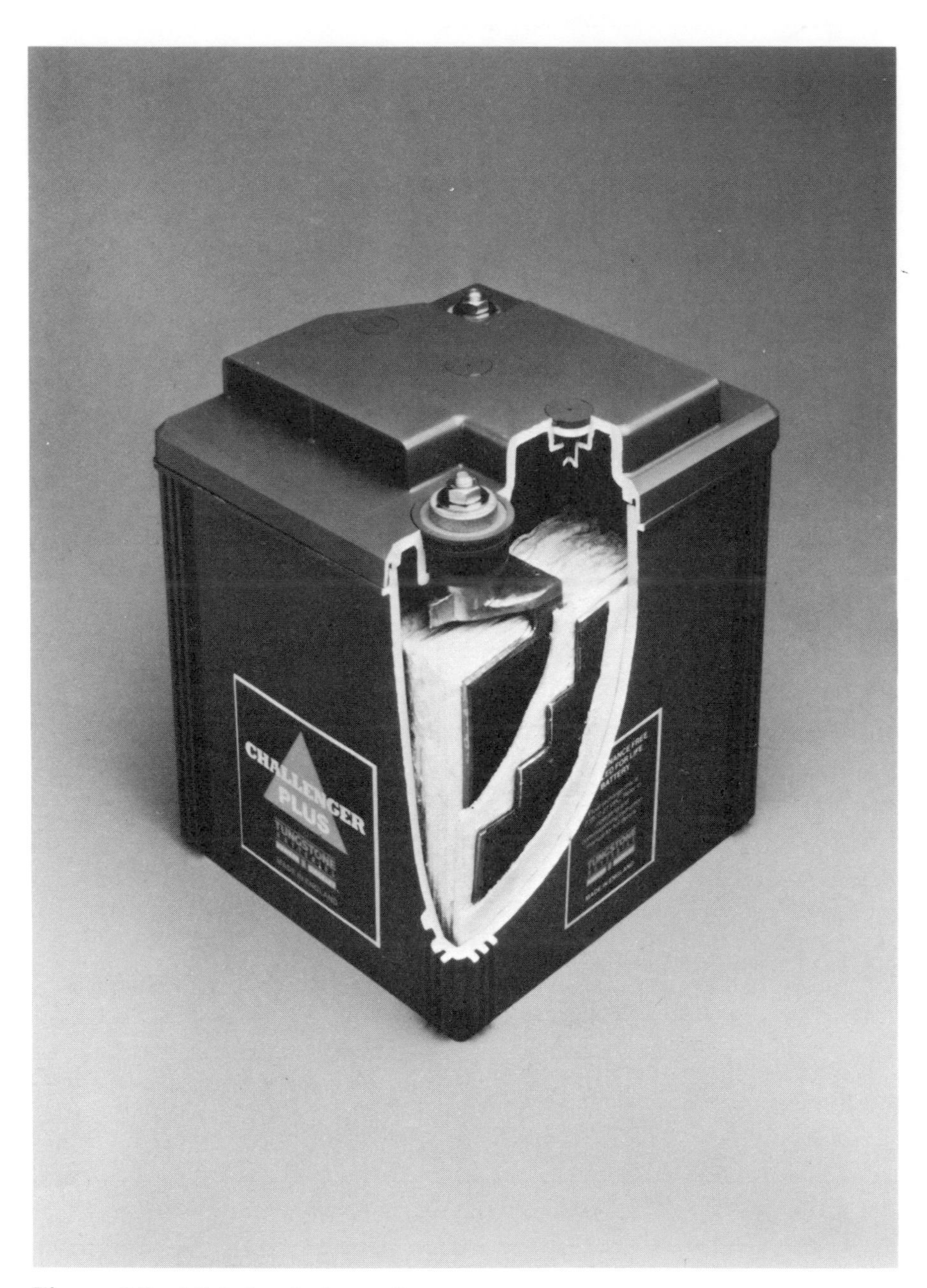

Figure 3.8 *Nickel-cadmium cell*

Alkaline cells

Alkaline refers to the electrolyte, which is mainly a solution of potassium hydroxide with the addition of lithium hydroxide. In the popular nickel-cadmium type (*Figure 3.8*), positive and negative plates are of similarly constructed fine perforated strips. Nickel hydrate is used for the positive plate with a conducting mixture of pure graphite. The negative plate consists of an admixture of a special oxide of iron with cadmium oxide. The chemical reaction of charging and discharging transfers oxygen from one set of plates to the other without affecting the chemical composition of the electrolyte, in sharp contrast to the lead-acid cell. A fully charged battery is at a high degree of oxidation and the negative material is reduced to pure cadmium. On discharge, the nickel hydrate is reduced to a lower degree of oxidation, thereby oxidizing the cadmium plate.

While the voltage per cell is lower than that for the lead-acid type, and generally the first cost is greater, it has the advantage of robustness due to the stronger plate construction and a steel plate housing. Also, charges are held for long periods and the cells can undergo charging and discharging with little damage.

The corresponding characteristics for the nickel-cadmium cell are as follows:

	EMF (V)	*Specific gravity*
charged cell	1.4	1.17
uncharged cell	1.2	1.27

Emergency lighting from batteries should be able to maintain the illumination from 1 to 3 h.

Sealed batteries

The sealed version constitutes a major advance in battery technology. Large batteries, with a life of at least ten years, are used in central systems. Self-contained luminaires use small cells which typically last about four years. The many other advantages claimed are:

1 no topping up of the electrolyte is necessary

2 clean working operation and protection is given by a welded-seal lid
3 the need for a separate battery room is eliminated
4 there is a good high rate discharge performance
5 they can withstand arduous conditions
6 the microfine glass mat retains the maximum amount of electrolyte in cells at all times.

Thus it can be seen that sealed batteries are ideally suited for emergency lighting. Note that lead cells, once charged (filled with electrolyte), must be kept charged, but Ni/Cd cells can be left in the discharged state for many months. For this reason, emergency lighting luminaires are often fitted with Ni/Cd cells.

Charging

The basic circuit may consist of a variable tapping transformer (*Figure 3.9*) to permit appropriate charging voltages. Since the current must be DC (with its output positive connected to the battery positive connection and the negative similarly connected) some form of rectifier is essential.

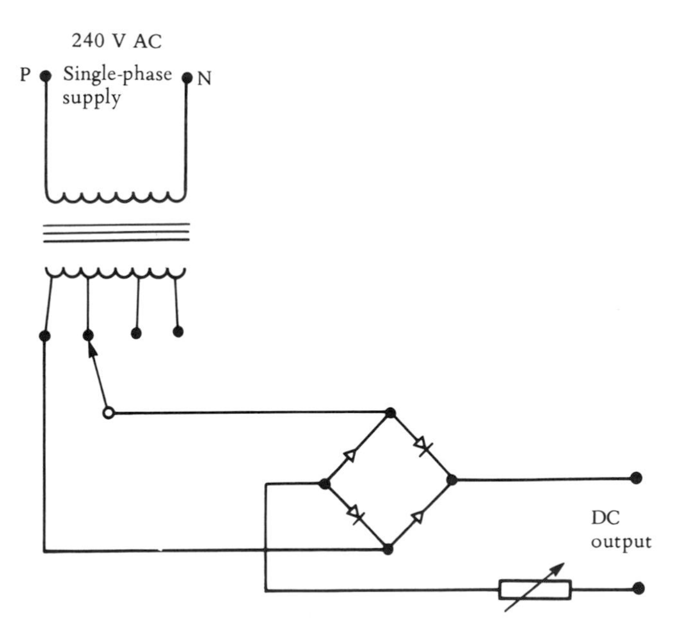

Figure 3.9 *Basic charging circuit*

Full-wave rectification by diodes as shown in the diagram is preferable. The rectified DC must be higher than the open-circuit voltage of the battery, i.e. 2.5 V–2.7 V per cell and approximately 1.6 V for the lead-acid and nickel-cadmium types, respectively.

To prevent the possibility of the current reversing in direction should the battery EMF rise to higher than that of the supply output, it is usual to fit a reverse-current cut-out.

Various methods of charging are:
1 *Constant current*. As the battery voltage rises during charge, the value of the series variable resistance shown in *Figure 3.9* is reduced in order to increase the charging voltage (alternatively this may be effected by operating the variable transformer tappings). Controls are manual or automatic.
2 *Constant voltage*. A voltage of 2.3 V–2.4 V per cell for lead-acid batteries is applied directly to the battery. (Checking is necessary as each battery has its own instruction book.) The battery EMF will be low at the beginning of the charge and a heavy current will flow. As the battery EMF increases, the current will fall. The circuit resistance prevents too heavy a current at the commencement.
3 *Trickle-charging*. This method is of particular importance for standby plant. The trickle-charging rate is about 1 mA per ampere-hour (Ah) of capacity for lead-acid cells up to 100 Ah. The continuous maintenance of this very low current keeps the battery fully charged without any gassing.
4 *Floating battery*. Statutory regulations stipulate that a secondary source of supply for emergency lighting must be available in theatres, cinemas and certain other public places. This has led to the floating system whereby the storage battery, although continuously in circuit, is trickled-charged. Thus the battery is kept fully charged and in the event of a mains failure it is automatically ready for essential standby services.

There are proprietary brands of automatic chargers which operate by continuously monitoring the changes in the battery voltage. When the battery is fully charged the control circuitry reduces the charging current automatically to a level sufficient only to maintain the battery characteristics.

Apart from voltmeter readings the state of the charge is obtained by a hydrometer showing the specific gravity of the electrolyte. The nozzle is dipped into the electrolyte and a small sample obtained by squeezing and releasing the top rubber bulb. The specific gravity of the lead-acid cell will probably lie between 1.280 when fully charged and 1.150 on discharge (these values are pronounced as 'twelve eighty' and 'eleven fifty', respectively). For exact values manufacturers' instruction sheets should be consulted.

4

Emergency lighting

Terms

Emergency lighting. This is simply stated by BS 5266 as 'Lighting provided for use when the supply to the normal lighting fails'. It is elaborated in the Chartered Institution of Building Services Engineers (CIBSE) technical memorandum T12 to 'Lighting provided for use when the supply to the normal lighting fails, in order to allow occupants to leave the place safely or to terminate hazardous processes before vacating the area'.

In the past appropriate exit signs applied to cinemas and theatres only. Emergency lighting is now also installed in hotels, blocks of flats, restaurants and other places where the public have access. Since the 1971 Fire Precaution Act this has been a requirement.

Escape lighting

Escape lighting is that part of emergency lighting which is provided to ensure that the means of escape can be safely and effectively used at all material times. (Note that these terms have certain features in common; it is therefore perhaps not surprising to observe that in commercial literature the definitions are often interchanged.)

Combined emergency lighting luminaire (sustained luminaire)

This consists of luminaires each containing two or more lamps, at least one of which is energized from the emergency supply and the remainder from normal mains. Such luminaires sustain illumination at all material times.

Standby lighting

Is that part of the emergency lighting that provides illuminance for certain activities to continue after failure of the normal supply. It may be fed from a generator or battery. The latter may include a constant-voltage charger. Fitting an auto/recharge facility ensures a fully automatic recharge. Standby lighting is often provided in shops or stores at checkout areas to illuminate the tills for security.

Maintained lighting

This type of emergency lighting is on the whole time that the building is occupied and consists of a distinct installation—wiring, luminaires etc.

Non-maintained lighting

Here the emergency lights are on only when the main lighting fails.

The various types of emergency lighting are not hard and fast. While the centralized system is essentially maintained, it may also be adopted for non-maintained lighting. At the same time, while single-point self-contained emergency luminaires are primarily fitted for non-maintained lighting, they are often employed as the maintained system for exit sign lighting.

General considerations

Statutory regulations insist on a secondary source of supply for emergency lighting in theatres, cinemas and certain other public buildings. Under these conditions maintained lighting is essential as it continually monitors the emergency lighting system and also acts as safety lighting for management control of the auditorium.

Emergency and standby plants are not new requirements. Their use in the past has been largely neglected, except for odd lamps dotted here and there in escape passages. Little regard was paid to the many strict requirements such as light levels, positioning, regulations (including British Standards), automatic activation or real allowance for emergencies.

Effects of emergencies. The conditions cover a wide range from minor shock or hurt to fatal injuries. The effect on machinery and electronic devices forming the basis of much of modern business and industry must also be taken into account. Now and in the foreseeable future its due importance needs to be appreciated.

Contractors, managers, engineers and technicians should be as knowledgeable as possible concerning these aspects. Alternative supplies of electricity will become necessary to keep industry running. At all times they are valuable assets to counter power cuts. Examples are hospital operating theatres and blood banks. In addition many working processes such as computer memory banks are inevitably affected by this type of breakdown.

The secondary lighting system, in addition to other requirements, must provide enough light to allow escape in an emergency along escalators and stairways. Generally it is accepted that the emergency illumination must not fall below 0.2 lux—although we will see that there are modifications—and should come on within 5 s. These values should be improved wherever possible. In the maintained system illumination will be continuous, irrespective of an emergency.

A sudden cut in illumination often leads to panic and emergency lighting is designed to prevent accidents which inevitably occur under such situations, particularly when there is the possibility of the outbreak of fire. Thus appropriate schemes must also fulfil the provisions of the statutory Fire Precautions Act 1971. Approval of the local fire authorities must be obtained before any work is carried out.

Central systems

They are not exclusive to, but generally follow, the central battery arrangement. In spite of the popularity of individual self-contained emergency luminaires, centralization has much to offer for large factories and commercial premises, especially where there are short cable runs. A variety of main unit sizes is available, often being housed in a sheet steel cubicle with a finish of corrosion-resistant stove enamel glass epoxy coating. Typical dimensions of the box shape (*Figure 4.1*) are width 0.8 m, length 0.6 m and height 1.7 m.

Mini sizes are suitable for situations such as public houses, small hotels, indoor sports and discotheques. Cubicles are available housing battery, instruments and control circuits for maintained and non-maintained lighting. Requirements may also be made for sustained lighting forms. The central battery feeds a number of outgoing slave luminaires wired in parallel. At the present stage of development manufacturers are in a position to offer tailor-made central systems suitable for particular establishments.

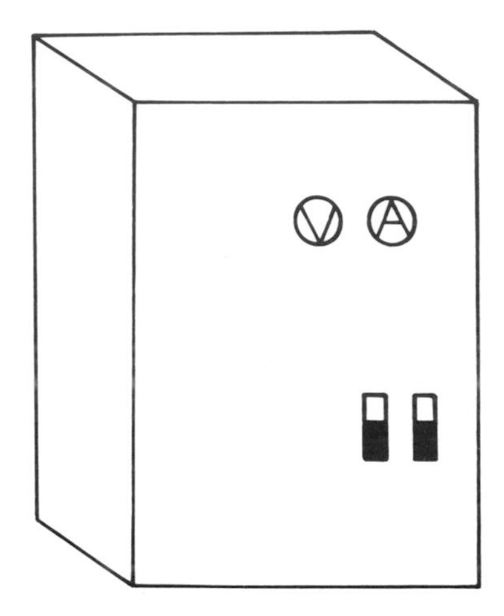

Figure 4.1 *Central battery cubicle*

One type of security lighting consists of an individual transistorized automatic 2-stage constant-voltage charger and changeover contacts. Under normal conditions, i.e. non-emergency, supply is derived from the 240 V AC mains and the battery is said to be floating while the inverter is running off-load. On failure of the mains supply a contactor changes over; the load is then connected to the inverter performing the function of charging DC from the battery to 240 V 50 Hz. With the restoration of the mains, the contactor transfers the load back to the mains supply and the battery is automatically recharged within 24 hours, inverter output being continuously monitored with provision for remote indication and alarm.

Alternatively, the battery may be fed from a diesel-driven generator, but allowance must be made for noise in running and exhaust fumes. There may also be difficulties in starting—a factor crucial to emergency lighting.

Choice of lead-acid types

Selection from three types may be made for central battery systems, largely dependent on the kind of plates used in manufacture.

Gent's offer a simple flat-plate battery giving a ten- to twelve-year life, with the advantage of a high capacity-to-volume ratio at a relative low cost. A transparent casing permits ease in checking the level of the electrolyte.

The tubular-plate type is another economical method of providing standby power. They are suitable for the kind of application that requires frequent charging and discharging. Each cell is also housed in a container made from styrene acrylonitrile material. Useful life is somewhat similar to the flat-plate battery. The positive plate is made up of lead alloy spines surrounded by fibre tubes filled with a mixture of lead oxides.

Where first cost is not a main consideration choice would go to the plate type. The positive plate has the characteristic of providing full rating throughout its life, since new active material is generated as the cell ages. The positive plate is formed to produce large (as much as twelve times) surface area. The negative plate is made by forcing an oxide paste into a cast iron alloy grid. Plates are interleaved with insulation to prevent any possibility of short-circuits from occurring. Plate separators should follow an amount as specified in BS 6290, Part 2. There is a long average life of 20 years, and some manufacturers, by special design, extend to 35 years with an ampere-hour (Ah) capacity up to 15 000.

Float-charging does not permit any interruption of the output supply. Here the battery, load and charger—of the constant-voltage type—are

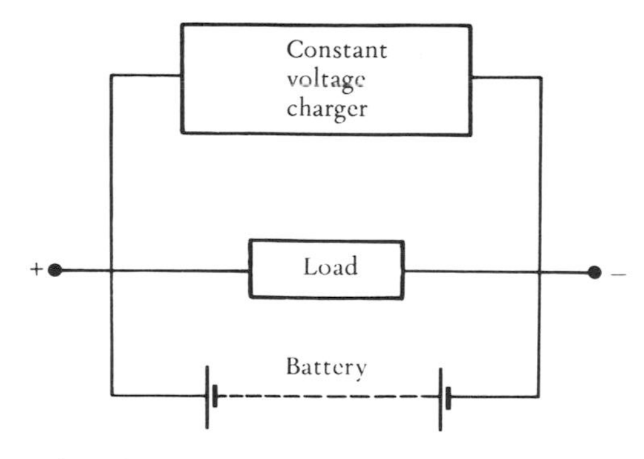

Figure 4.2 *Float-charging*

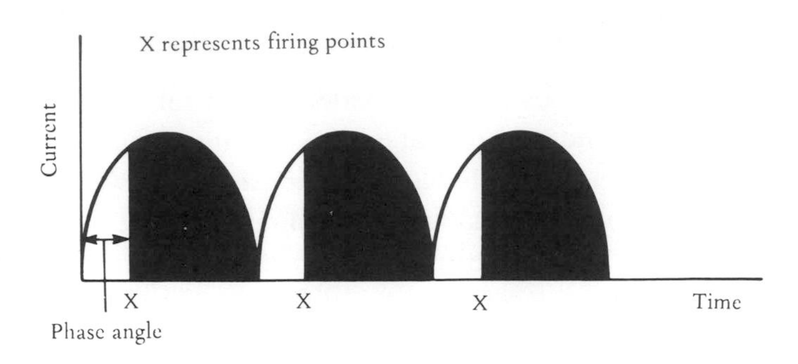

Figure 4.3 *Thyristor phase control*

connected in parallel (*Figure 4.2*). This type of charger consists of a statically controlled rectifier employing a thyristor or similar device to maintain output. The thyristor acts by variations in firing at particular angles of the AC wave (*Figure 4.3*). As an alternative the impedance of a transductor is made to vary the magnitude of saturation produced by DC in an auxiliary winding.

Waveforms

As part of a central system an inverter may possess a sine-wave or square-wave output. Both types are used to operate tungsten or fluorescent lamp luminaires. Square waves have a high efficiency, thus providing economic running, but generally must be used with power factor correction capacitors. Originally they were designed to improve a low harmonic, low distortion synthetized sine wave output.

Slave luminaires

Care must be taken for the prevention of slave lamps operating by energy control devices fitted between luminaires and the central emergency supply. An inverter may be necessary for slave fluorescent units. They should be designed and constructed to the requirements of BS 4533, Part 102.22. Safeguards are offered against the entry of foreign bodies by the appropriate IP classification. It will provide dust-tight/jet-proof protection which can be achieved with an acrylic diffuser, vandal-resistant need polycarbonate.

For large premises, a number of central battery controls—say, one on each floor—to feed the slave luminaires would make for additional emergency lighting protection. They can be supplied to operate on AC/DC 24 V, 50 V or 110 V alternative maintained or non-maintained systems. Manufacturers offer a wide range of slave luminaire finishes from plain non-corrodible heavy-gauge aluminium types to attractive sparkling crystal glass, surface or recessed mounting. Compatibility is required between the slave luminaires and the central system.

Maintained system

The luminaires are operated during the whole of the material period that the premises are in operation. In the centralized arrangement the generator may feed the emergency light directly or through charging a battery, which in turn supplies the secondary installation. Again, referring to the centralized method, the wiring, luminaires etc. form a distinct installation. This method offers considerable advantages for large buildings, stores and constructional sites where the emergency installation could be fed from a diesel-powered generator. Performance and maintenance would be affected by the size and shape of the structure. The engine can be fixed directly on to a concrete bed by resilient mountings. It may, however, be preferable to mount it in a raised position by resting it on solid plinths, thus allowing easy access for adjustments, repairs and ease of maintenance. Raising also leaves room for a drip tray.

'Noiseless' small portable generators possess attractive features. Alternatively, correctly fitted baffles reduce, or even may entirely eliminate, this airborne nuisance. Diesel oil storage is normally subject to local fire regulations.

Additionally, IEE Wiring Regulation 422–5 stipulates that, where the machine has a capacity of more than 25 litres, means must be provided for draining away surplus oil in any other part of the building. In a note to the Regulation, further precautions are given:

1 a drainage pit 'to collect leakages of liquid and their extinction in the event of a fire', or

2 'installation of the equipment in a chamber of adequate fire resistance and the provision of sills or other means of preventing burning liquid

spreading to other parts of the building, such chamber being ventilated solely to the external atmosphere'.

With respect to the growing realization of the deleterious effects of burns, a succeeding Regulation stipulates that, should any part of electrical fired equipment reach a temperature of more than 80°C, means must be adopted to prevent accidental contact with such equipment.

Non-maintained system

In this type of operation the emergency lamps are energized only on failure of the normal lighting, usually by an automatic changeover switch. The simplified circuit diagram shown in *Figure 4.4* indicates how a cut in the mains operates a relay, automatically switching on the emergency supply.

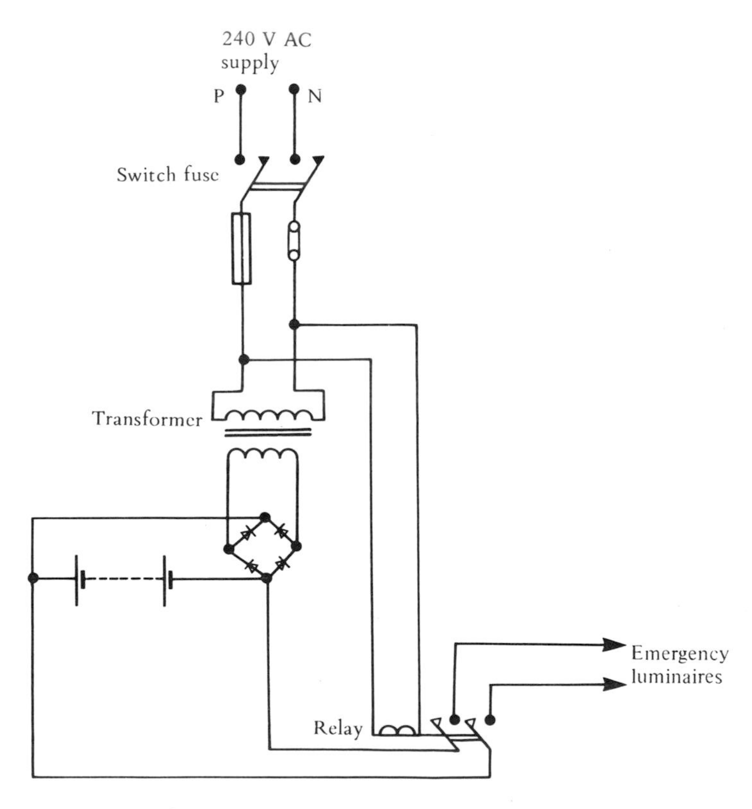

Figure 4.4 *Wiring for centralized non-maintained emergency lighting system*

Single-point units

As a complete contrast to the centralized system, battery-operated individual units house one or more fluorescent tubes, with a trend towards high-frequency lamps, and contain a sealed nickel-cadmium cell, constant current charger and a solid-state changeover switch (*Figure 4.5*). A light-emitting diode (LED) indicates that the battery is connected and is being charged. In the event of a mains failure, instant, i.e. within 5 s, emergency lighting is provided.

Non-maintained, maintained and sustained emergency lighting indicate the wide range of single-point units that are available from manufacturers. Sustained emergency lighting represents a luminaire which contains two or more light sources, one of which will only illuminate in the event of a mains failure, the other tube(s) being energized from the normal mains supply.

For ceiling mounting, the control gear could be concealed within a false ceiling or void. Large luminaires permit the gear to be fitted within the luminaire itself. Care is necessary to guard against excessive heat, so that in this case the equipment must be installed remote from other heat sources.

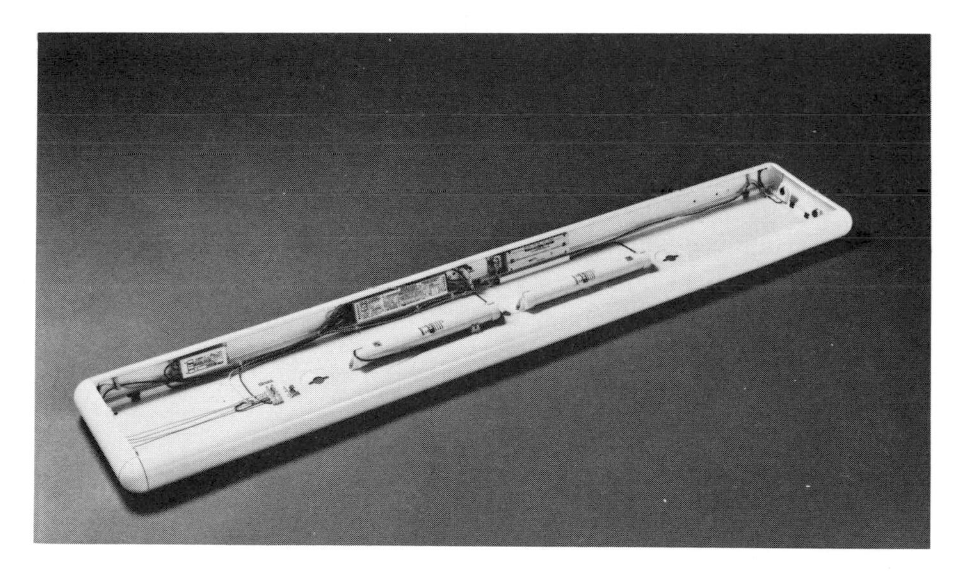

Figure 4.5 *Integrated emergency light unit*

For self-contained luminaires, when the changeover switch is operated to give an emergency supply from the charged battery, illumination from the battery may give a minimum time supply of 3 h. However, TM15 cites 1 h as sufficient for orderly evacuation of the premises.

Self-contained conversion kits

Emergency units, as previously described, whether centralized or single point, require a completely fresh installation. Economies can be obtained by conversion packs which are able to modify existing fluorescent luminaires for use as emergency lighting units. Typical self-contained conversion kits comprise a lampholder reflector assembly and separate gearbox containing a module and battery pack to use with standard surface-mounted or recessed luminaires to provide a self-contained non-maintained emergency lighting facility of 3 h duration. Equipped with a 300 mm, 8 W white fluorescent tube, the reflector assembly is complete with an LED charge indicator. Others convert up to an 1800 mm 70/85 W lamp. The control unit is fitted remotely, or within the spine of the luminaire, and if mounted in a void it should not be mounted on top of the luminaire or blanketed by other fitments.

Some arrangements convert existing fluorescent luminaires into maintained emergency lighting units. They operate with a single tube fitting or one tube of a multiple tube luminaire. In the event of a mains failure, the relevent tube will be illuminated. On restoration of the mains supply, the tube fed from the mains will resume normal operation and the battery will be recharged by the inverter/changeover unit. Emergency illumination will not fall below 0.2 lux, in conformity with BS 5266.

A useful practical point to be noted is that with the larger powered lamps a high level of illuminance is achieved, and consequently the number can be reduced compared to the lower wattage tubes, provided the luminaires can distribute the lumens over a wide area.

Which system?

Advantages of central battery over single-point arrangement:

1 cost per point reduced for large areas

2 facilitates maintenance and monitoring of lamps
3 quality product will ensure long life; manufacturers will cite 25 years

Disadvantages:

1 very careful assessment essential for possible later extensions
2 wasteful if supply unit is too large and costly if later extension necessitates a larger unit
3 battery failure will put all emergency lamps out of action
4 distinct installation required
5 heavier cables essential for long runs in order to minimize voltage drop.

ICEL

As previously mentioned, the Industry Committee for Emergency Lighting (ICEL) exists to help and guide users, consultants, contractors and members of the electrical installation industry in all matters which touch on all forms of emergency lighting. It is particularly concerned with contributing to national and international standards matters, e.g. the British Standards Institution (BSI), the International Electrotechnical Committee (IEC) and the European Committee for Electrotechnical Standardisation (CENELEC), to ensure that the UK market has the latest and best technology available as quickly as possible.

As mentioned in Section 1.7, ICEL was originally formed in 1978 from the emergency lighting sections of the British Electrical and Allied Manufacturers' Association (BEAMA) and the Lighting Industry Federation (LIF). It was formed as a joint industry committee in response to a need to formulate and promote industry standards for emergency lighting equipment based on considerable experience.

The ICEL have now produced standards for the construction of central battery, emergency and self-contained system complete with comprehensive guides to their application. Compliance is shown by the ICEL certification (*Figure 1.15*). A relevant ICEL publication (ICEL 1001) refers to as many as twelve British Standards. In the second edition of ICEL 1001 Part 2 coverage is given to self-contained luminaires and associated equipment. It states that these luminaires are suitable for mounting on normal flammable surfaces when holding the 'F' mark. In addition, onerous tests for endurance and thermal conditions are detailed, and reference is made to the IEC requirements for DC supplying electronic ballasts.

ICEL 1001 Part 3 deals with electric lamps essential to emergency lighting and the ICEL 1003: 1982 publication is headed 'Emergency lighting'.

Applications guide: ICEL: 1004: 1979 serves as the 'Industry guide to the use of emergency lighting modification units/conversion kits'. ICEL 1005 sets out further test devices for luminaires. Thus it can be seen that the second edition of ICEL complements the current edition BS 5266 for emergency lighting.

Escape routing

The tragic King's Cross Station fire (1987) is a vivid reminder of the necessity for proper escape routes. Technical Memorandum TM 12 as produced by the Chartered Institution of Building Services Engineers (CIBSE) serves as a valuable guide to the basic emergency lighting standards BS 5266 and BS 5225: Part 3: 1982 'Method of photometric measurement of battery-operated emergency lighting luminaires'. TM 12 gives a similar definition to BS 5266 for the route which emergency lighting should follow, i.e. the escape route, as 'A route forming part of the means of escape from a point in the building to the final EXIT'.

Panic and confusion are likely to occur following a sudden blackout in a building inhabited by many people. Dangers from the outbreak of fire and chemical exhausts are among the other major hazards which a clearly defined proper escape route can minimize.

BS 5266 cites the positioning of emergency lighting luminaires along escape routes: luminaires should be placed at each EXIT and emergency EXIT door and at other points where there may be potential hazards, e.g.

1 near each intersection of corridors
2 near each change of direction
3 at staircases; in particular each step to receive direct light, thereby avoiding shadows. This siting also applies to changes in floor level
4 at each final EXIT.

The aim is to avoid collisions and thereby lessen the possibility of people becoming panic-stricken, in addition to assisting rapid evacuation of the premises.

One possible arrangement is set out by Chloride Systems (*Figure 4.6*). By now it will be appreciated that rarely are two layouts the same, owing to

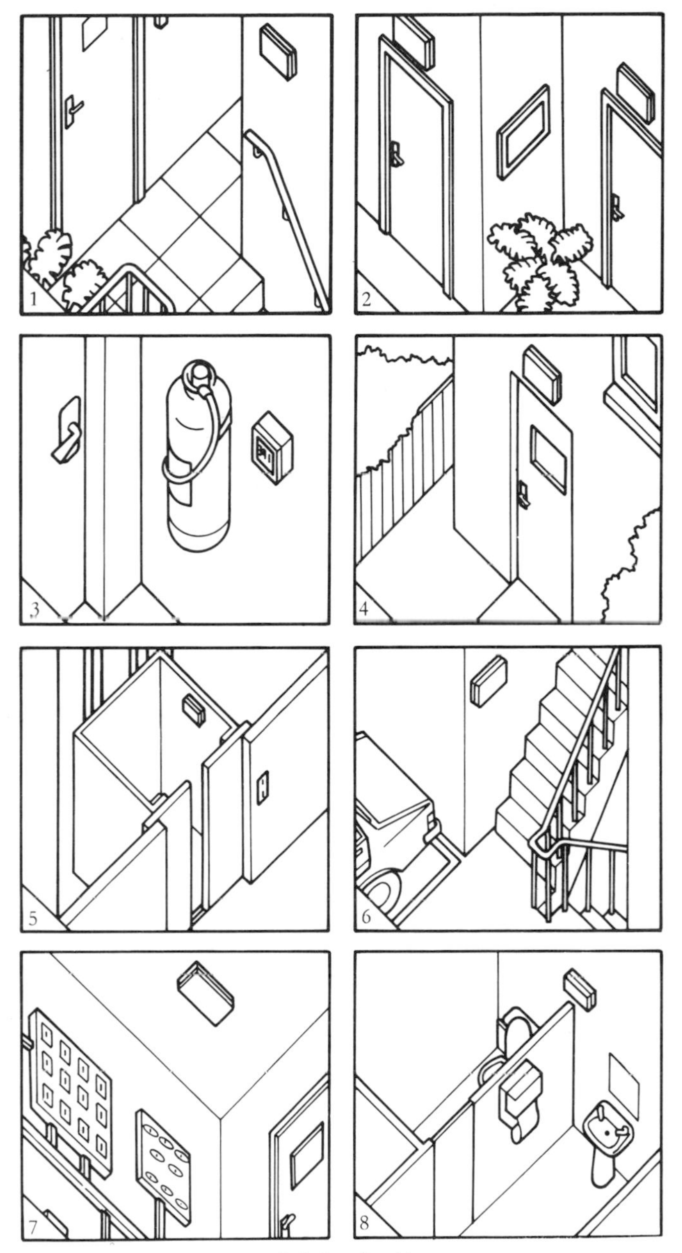

Emergency lighting should cover:

1 Changes of direction.
2 Final exits.
3 Fire protection equipment.
4 Outside final exits.
5 Lift cars.
6 Car parks.
7 Plant rooms.
8 Large toilets.

Figure 4.6 *Positioning of emergency luminaires*

differences in building construction and the amount which the owners of premises are prepared to spend. However, all schemes must be ready for instant operation of the secondary lighting in the event of an emergency and clearly indicate escape routes. The horizontal illuminance on the floor centre line of a clearly defined escape route, which can be measured by a photometer, must not fall below 0.2 lux. This value of illuminance also applies to any part of the floor where there is no clearly defined escape route, such as may occur in open-plan offices. However, escape route signs must be visible from all parts of the area.

EXIT signs should be fitted at a height between 2 m and 2.5 m from the floor level—luminaires may be obscured if placed too low, or they may be out of the line of vision if mounted too high. The deleterious effects of glare and dazzle are two other considerations which must be taken into account.

Distances between luminaires in long corridors and passages require information as supplied from lighting manufacturers' data and tables. This information is probably obtained from isolux diagrams and/or computer analysis of photometric material. With fluorescent emergency luminaires the following siting will also need to be taken into account (see *Figure 4.7*):

1 transverse distance to wall
2 transverse to transverse spacing
3 spacing axial to axial
4 axial distance to wall.

Further positions for luminaires are:

1 passenger lifts
2 moving stairs and walkways, even if not part of the escape route
3 toilet accommodation where floor area is in excess of 8 m^2.

Wiring

As required in security wiring, all cabling and equipment is to be installed, worked and maintained in a safe and efficient condition at all times. Thus the work and materials must be of a particularly high standard. Permitted wiring systems are:

1 mineral insulated metal sheathed cables made to BS 6207. Under

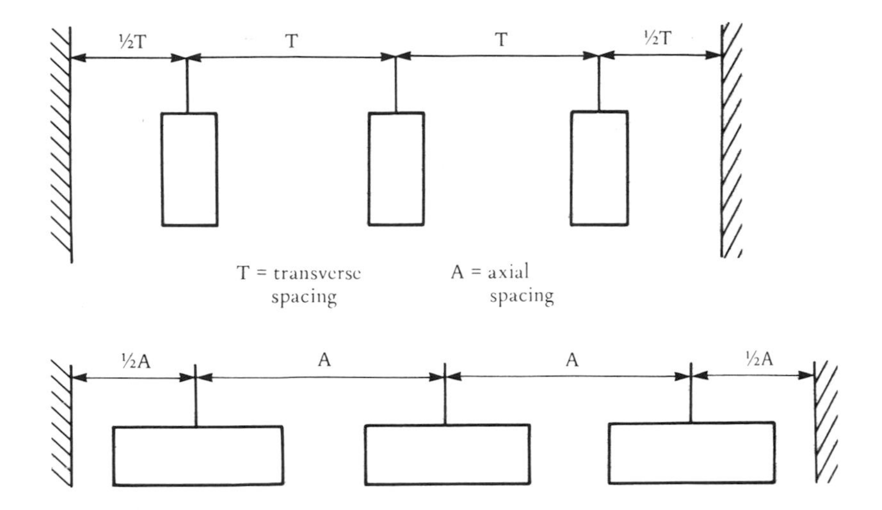

Typical distances (m)

T	A
16	10
17	11

Figure 4.7 *Emergency luminaire spacing*

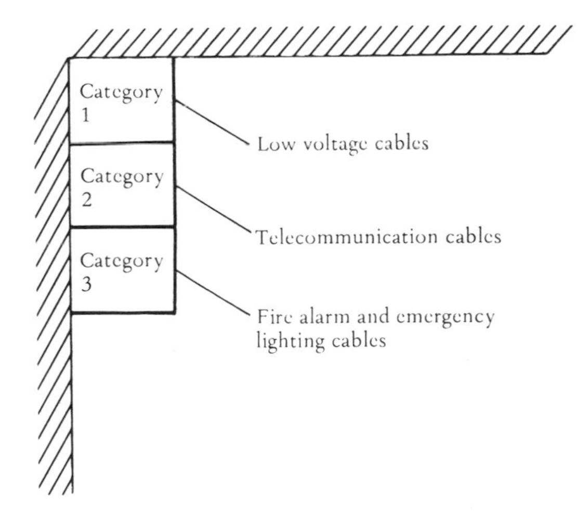

Figure 4.8 *Multiservice cornice trunking*

possible damp or corrosive conditions, cables should be PVC-sheathed and glands fitted with PVC shrouds

2 wire-armoured or aluminium strip-armoured cables to BS 6346
3 elastomer insulated cables complying with BS 6007
4 PVC-insulated cables to BS 6004 housed in conduits. While 1 mm^2 is the minimum cross-sectional area, a larger size will usually be required in order to minimize the effects of voltage drop.

Segregation

The IEE Wiring Regulations stipulate three categories of circuit which require segregation into separate rigid divisions of cable trunking (*Figure 4.8*) to prevent physical contact between the various categories.

Category 1: Mains supply voltage.

Category 2: Radio, telephone, sound distribution, intruder alarm, bell and call and data transmission circuits as supplied by one of the following safety sources:

(a) A BS 3535 Class II safety isolating transformer, the secondary of which is isolated from earth. (Class II equipment safeguards against electric shock by not relying only on primary insulation; a supplementary insulation is provided; no provision is made for connection to exposed metalwork by a protective conductor.)

Category 3: Fire and emergency lighting circuits.

There are a number of exemptions. Should Category 2 cables have insulation which is equal in all respects to Category 1 cables, then they may be contained in the same trunking division. This requirement also applies to multicore cables containing Category 1 and Category 2 cables. Again, while fire alarm cables of the mineral-insulated type are permitted to be contained in the same channel as Category 1, it is preferable to place all fire alarm cables in a separate division. Under no circumstances may Category 1 and Category 3 cables circuits be contained in the same multicore cable or flexible cord.

Where Category 1 and Category 2 cables are taken into a single outlet, then a partition of rigid fire-resisting material must be positioned between the two category cables.

When referring to equal insulation and various voltages, account must be taken of the extra-low-voltage (ELV) and low-voltage ranges (LV);

(ELV is the spread of 0–50 V AC or 0–120 V DC. LV covers the range of AC circuits (RMS values)):

DC { 50–1000 V between conductors or
50–600 V between conductors and earth

AC { 120–1500 V between conductors or
120–900 V between any conductor and earth (*Figure 4.9*)

Generally, ELV circuits must be segregated from LV circuits.

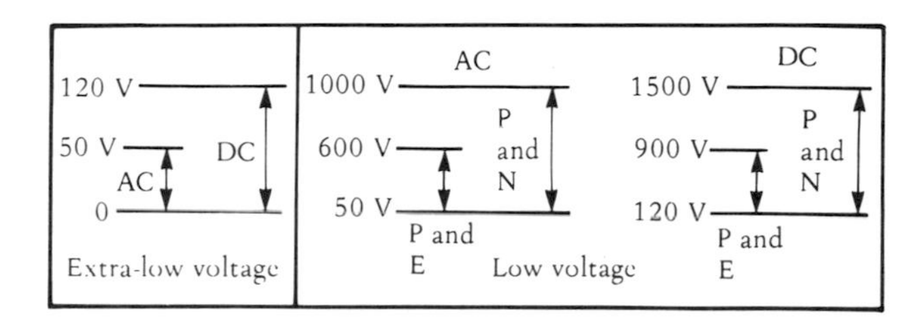

Figure 4.9 *Voltage ranges*

Although the self-contained luminaires constitute one form of emergency lighting, rather surprisingly the circuits supplying them are not classified as emergency lighting circuits. To accord with this classification the supply to emergency lighting luminaires must form a distinct circuit or be slave luminaires fed from one or other safety source as follows:

1 generator so designed as to attain the necessary illumination level within 5 s (or 15 s in certain circumstances). The generator may supply the emergency lighting circuit directly or via a battery charged by the generator
2 centrally located rechargeable battery
3 a class II isolating transformer to comply with BS 3535 with the secondary winding isolated from earth
4 a supply providing equivalent protection to that of a safety transformer
5 an electronic device with sufficient safeguards so that if there is an internal fault the outgoing voltage will not exceed ELV.

Note: The regulations for safety sources are outside the scope of the IEE Wiring Regulations. Neither do they apply to explosive hazards such as are stated in BS 5345, 'Selection, installation and maintenance of electrical apparatus for use in potentially explosive atmospheres (other than mining applications or explosive processing and manufacture)'.

Design features

Certain criteria have already been stated. Wiring, controls and luminaires are expected to be of the highest grade. The inclusion of standby lighting allows certain work operations to be unaffected even during the event of a mains failure. This will be of prime importance in permitting reoccupation of premises after initial evacuation, to check thoroughly whether any of the personnel or key materials have been left behind. Allowance must also be made so that the emergency lighting is on for at least 1–3 h. At the planning stage ease of maintenance requires consideration; this also applies to the purchase of material and the replacement of spare parts. Thus, although foreign equipment may be cheaper at first cost, the availability of spare parts must also be taken into consideration.

Thought must also be given to anticipate circumstances which may arise. As an example, the practical engineer will plan to offset the deleterious effect of heat on secondary cells. Costly extensions can be lessened by foresight in making appropriate provision at an early stage.

5

Security lighting

Burglar alarms

Closed-circuit wiring is advisable, as with this method any break or cutting of the wires sets off the alarm bell. A traditional example is illustrated in *Figure 5.1*. The ball contacts (which may be replaced by magnetic catches) are embedded in door jambs and window frames; contacts may also open pressure mats. The day switch allows the system to be cut out during the period of normal operation. The time delay switch ensures that a responsible person may leave the premises with the alarm circuit operational and fully working.

Mains supply is rarely adopted because of the possibility of supply cuts during a critical period. Even trickle-charging may not be favoured due to the possibility of a breakdown. In this respect manufacturers of burglar alarm equipment often recommend long-life dry primary cells, which should be renewed at definite periods.

Photoelectric apparatus can be adopted to give protection across doorways, access to safes and at other strategic positions. In addition, the breaking of an invisible infra-red beam in conjunction with appropriate relays can be made to set off alarms. Light-sensitive semiconductors may also be employed to have the same effect.

Light against crime

While an internal closed-circuit burglar alarm has great merits, its effectiveness would be improved by appropriate external, and sometimes

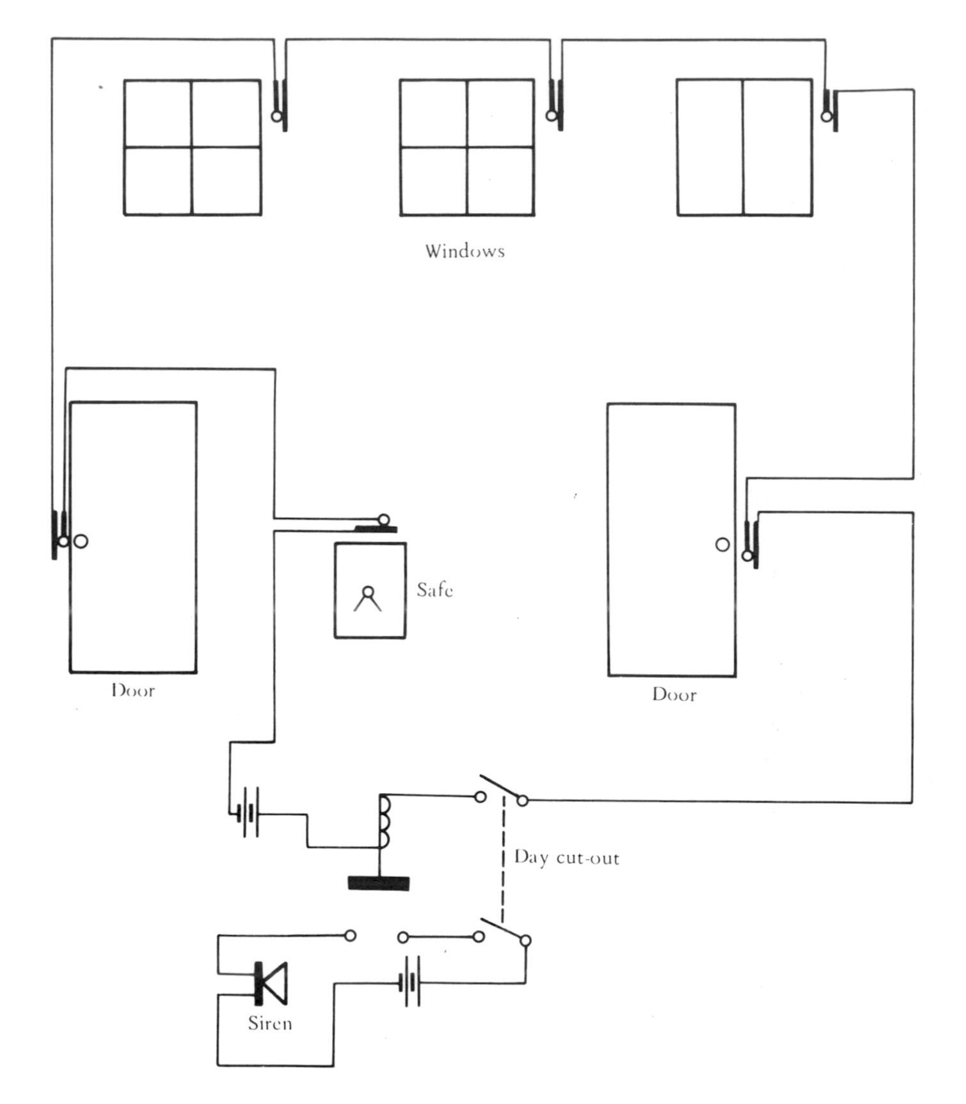

Figure 5.1 *Traditional burglar closed-circuit*

internal, illumination. This specialized form of lighting is claimed by police, insurance and government experts, to act as a formidable deterrent to the would-be intruder. There is an authoritative Home Office Research Bulletin which has an article entitled, 'The case for lighting as a means for preventing crime'. An extract states, 'Indeed the scope for guidance — and for the implementation of lighting programmes aimed at reducing crime — is enormous'.

These are lawless times. Vandalism, robbery and acts of violence are continually increasing in number. It is not always appreciated that mugging, in addition to physical injuries, may leave the victim with permanent psychological scars.

Crimes unfortunately show an upward trend. Typical values of losses in London alone are estimated to be increasing by £20 million annually. Taking the country as a whole, criminal losses are said to be mounting by more than £60 million a year. We appear to be reaching a stage in the UK when the number of forced entries will be at the rate of one per minute.

Industrial and commercial managements must realize that this is by no means the whole picture. These destructive acts bring in their train loss of production, delays causing a deterioration in customers' orders and goodwill, and may produce an enormous dislocation in essential services. Wilful damage to property is high in the list, including possible resultant outbreaks of fire. It is therefore not surprising that increased emphasis is being placed on security lighting. The criminal shuns light and prefers to carry out his nefarious activities under the cover of darkness. Lighting in multiple stores and open-plan offices during the hours of darkness gives a reasonable surveillance to guard personnel and police.

A major aim is to prevent the burglar from even reaching the front door. To safeguard people and property, the lighting must be designed so as to be installed in conjunction with

1 appropriate barriers
2 locks at doors and windows
3 locks at other access points
4 security personnel (*Figure 5.2*).

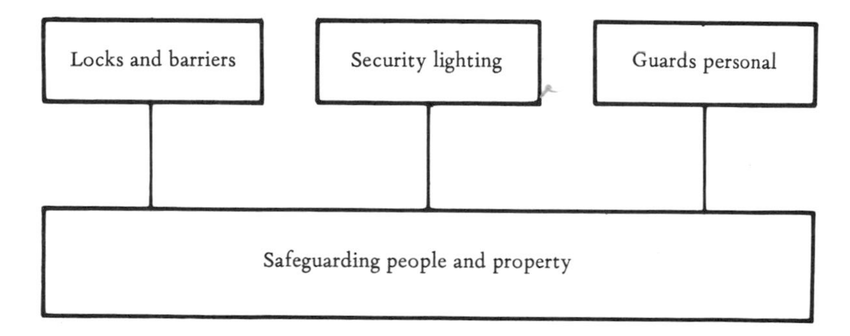

Figure 5.2 *Elements against unwarranted break-in*

The lamps for exterior lighting must be positioned so as not to cast any shadows which may serve as a cover for the would-be intruder. Linked perimeter chains are preferable to dwarf walls which may act as a shield.

On the positive side, the cost of security lighting can be somewhat offset by a reduction in insurance premiums, and it makes the task of police and guards easier in patrolling the premises; some premises also lend themselves to forming part of an emergency lighting system. Amenity lighting is a further advantage and may be obtained by arranging the security lighting to improve the appearance of the building structure. Picking out the aesthetic and architectural features often enhances the prestige of the firm and its products, thereby becoming an incidental visible aspect of advertisement.

The practising engineer or contractor specializing in lighting, assesses from experience the positioning, sizes and types of lamps required. Certainly, for the smaller installations the rule of thumb often produces the desired result at a reasonable cost. There are, however, situations where calculations are required: for floodlighting, in particular. Some basic examples are given in the floodlighting section.

Similarly to emergency lighting, large installations may be supplied by a generator with an appropriate battery and inverter.

Lamps and luminaires

Here we are dealing with light fitments which directly apply to security lighting, supplementing the general illumination principles as described in earlier sections.

Generally, all external lighting, contrasting to complete darkness, has a certain deterrent effect. The lighting must be arranged so as not to be controlled by unauthorized personnel. However, for maximum effect special design is essential. For example, purpose-made luminaires are often constructed of polycarbonate, which is a very strong plastic for the body part. They need not be too severe in appearance—with a modern look they may be very presentable.

When fitted externally, because of the vagaries of British climate they should be water- and corrosion-proof. They are sometimes erroneously described as ‘vandal-proof’, but this is not strictly true when taking into consideration the destruction caused by savage attacks for any length of

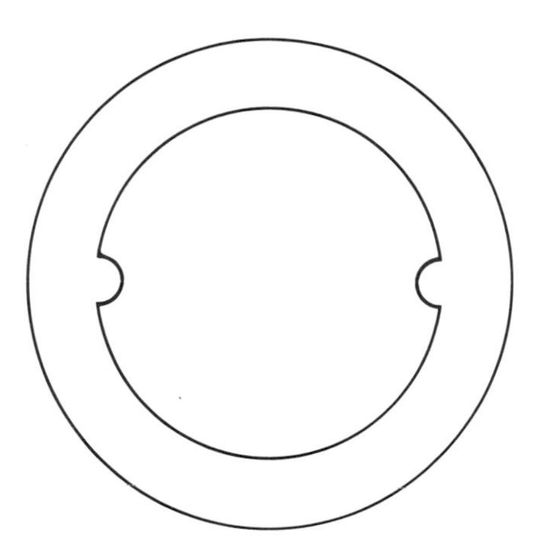

Figure 5.3 *'Secret' screwhead*

time. It has been stated that a rifle shot can be made to penetrate a sample of 18 gauge sheet steel at close range. Thus the present trend is to call external luminaires 'vandal-resistant'.

To add to the intruder's difficulties, the installer should consider unorthodox fixing methods by replacing the standard slotted or Philips screwhead by secret types. A possible example is illustrated in *Figure 5.3*.

There is a wide choice of security luminaires. Some give a simple forward light, while others have the advantage of additional side light. While not vandal-proof, they must be at least vandal-resistant. The present trend is to supply them complete with compact fluorescent lamps. *Figure 5.4* shows a security bulkhead, which may be wall- or ceiling-mounted, complying with BS 4533, Part 102.1 (IP 65 dust- and jet-proof), and can be supplied with a prismatic version of the 16 W bulkhead. 20 mm-threaded side entries are standard, together with flush-finish tamper-resistant screw-in blanking plugs. Provision at the back is made for BESA fixing or cable entry with alternative fixing centes. Well glass luminaires may be fitted into a right-angle swan-neck bracket (*Figure 5.5*), enabling the light to be thrown forward. Brackets with other angles are obtainable to suit particular circumstances.

Regular patrolling of large premises is expensive, and perhaps would be justified only for vital military establishments or similar important buildings. Security lighting round the building's perimeter provides a sound line of defence. This arrangement is sufficient to act as a powerful

Figure 5.4 *Bulkhead fitting*

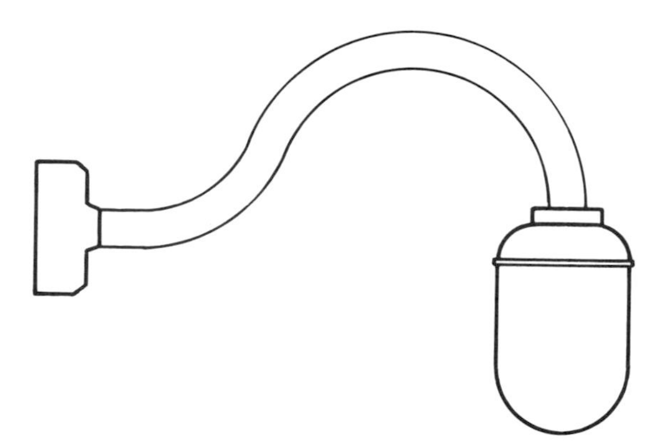

Figure 5.5 *Swan-neck well glass luminaire*

warning and deterrent to the wrongdoer, who must show himself by passing through a zone of light to gain entry.

The Electricity Council in their excellent publication, 'Essentials of Security Lighting', give a selection of lamps. An extract, by courtesy of the Electricity Council, is set out in *Table 5.1*.

From *Table 5.1* certain observations may be made.

The compact fluorescent lamp is subject to continuous developments. In general it consumes approximately a quarter of the power with burning hours five times as long as an equivalent GLS lamp, thereby offering, under suitable conditions, special advantages for emergency and security lighting. They fall into the following categories:

1 with a bayonet cap (BC) or edison screw (ES) cap they may be fitted as a direct replacement to GLS lamps, an electronic ballast, starter and radio suppressor being inserted in the base
2 for new luminaires the auxiliary gear is fitted separately and may be wired to two or more lamps
3 manufacturers list 7, 11, 15 and 20 W types as equivalent to the 40, 60, 75 and 100 W GLS lamps
4 they give good colour rendering with a quick starting operation.

Security lighting systems may come under the headings of class A, B and C.

Class A indicates an *extreme security risk* area requiring the lamp units to be so constructed as to be able to resist heavy damage. Where fitted to external columns, sturdy locks must be fixed to control gear compartments.

Table 5.1

SOXE low-pressure sodium lamps

Wattage range:	18 to 131
Efficacy:	67 to 165 lm/W
Life:	Circa 8000 h
Restrike:	Igniter circuits instantaneous while lamp is hot.

These lamps produce the most light for a given input and are very economical in use. They are not recommended where the best visibility is required at extreme range.

SON high-pressure sodium lamps

Wattage range:	50 to 1000
Efficacy:	55 to 110 lm/W
Life:	Most ratings in excess of 8000 h
Restrike:	Igniter circuits usually within one minute. Others 5 min or longer.

A general-purpose lamp with economic running costs. Other versions with higher output (SONP) and better colour rendering (SONDL) are available.

MBI metal halide lamps

Wattage range:	75 to 2000
Efficacy:	50 to 85 lm/W
Life:	Circa 7500 h depending on operating conditions
Restrike:	Igniter circuits usually within one minute. Others 4–5 min.

For use where colour rendering is of prime importance.

MBF colour-corrected mercury discharge lamps

Wattage range:	50 to 2000
Efficacy:	30 to 55 lm/W
Life:	7500 h
Restrike:	5 min or longer.

A cheaper circuit than MBI but now largely superseded for security purposes.

MCF tubular fluorescent lamps

Wattage range:	8 to 125
Efficacy:	20 to 75 lm/W (high-efficacy colours)
Life:	1.2 m and longer 7500h; others 5000h
Restrike:	Immediate.

A general-purpose lamp for short-range use. Very cold weather can significantly reduce light output.

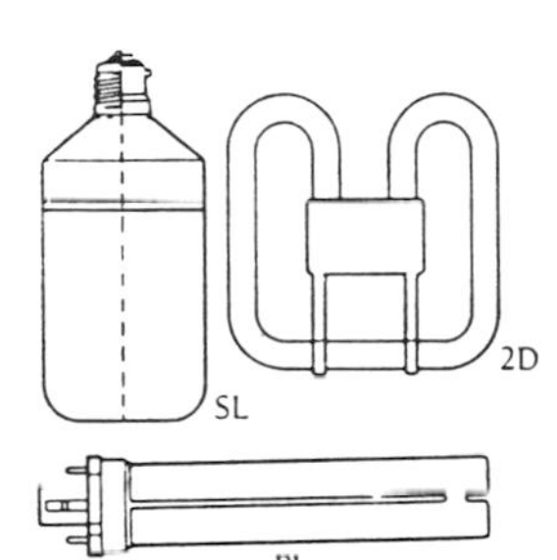

CFL compact fluorescent lamps

Wattage range:	7 to 35
Efficacy:	Circa 50 lm/W
Life:	5000 h
Restrike:	Immediate.

Available as original equipment but may also be used as a filament lamp replacement in small bulkheads etc.

TH tungsten-halogen lamps

Wattage range:	200 to 2000
Efficacy:	16 to 22 lm/W
Life:	2000 h
Restrike:	Instantaneous.

A floodlighting lamp for short- to medium-range distances. Its high energy consumption makes it uneconomic for use on a large scale.

GLS lamps

Wattage range:	40 to 2000
Efficacy:	10 to 18 lm/W
Life:	1000 h (some ratings are available in 2000 h or longer versions)
Restrike:	Instantaneous.

Smaller wattages useful for infill purposes, but short life and low efficacy make them uneconomic for large-scale use.

Similar to certain forms of emergency lighting, the electrical system should incorporate a standby service which will give a minimum of 2 h of battery supply to the more important lighting positions.

Class B—*high risk*. Here the installed luminaires must also be of a very heavy construction so as not to permit breakage by stone-throwing. Fittings which are automatically switched on as darkness falls by means of photoelectric cells come under this category. The special switchgear must be available only to authorized personnel, and could also, preferably, be linked to an alarm system.

Class C—*moderate risk*. The installation must be planned so as to avoid the possibility of tampering by intruders. The choice of lamps must include filament, tungsten halogen, fluorescent and sodium (high-intensity discharge HID) lamps.

Commercial premises

At one time offices held only stationery, paper files, typewriters and other items of little value to an intruder, so that it was considered barely worth while to spend money on security. With the advent of electronics into commercial and business premises with costly and sensitive data storage, computers, microprocessors and other expensive equipment, the situation has radically altered. Unfortunately such items of equipment when stolen are easily disposed of for cash.

Entry will almost certainly be made through doors or windows; therefore for maximum security they need to be well illuminated by carefully selected and placed luminaires.

Supervision of premises by a hand torch is a doubtful deterrent; it may scare off the amateur crook but does not deter the professional because it gives him the opportunity to seek concealment or even to break in at positions where the beam from the torch has not projected.

Patrolling should be at irregular times and generally follows varying routes. Wherever possible time-switches (*Figure 5.6*) require adjustment, so that the ON and OFF controls are changed from day to day. Men guarding the premises require proper training to be on the alert for any untoward event — movement or noise. Their work is greatly assisted by strategically positioned light units. The external lighting may require to be supplemented by internal office (especially if 'open plan') and store

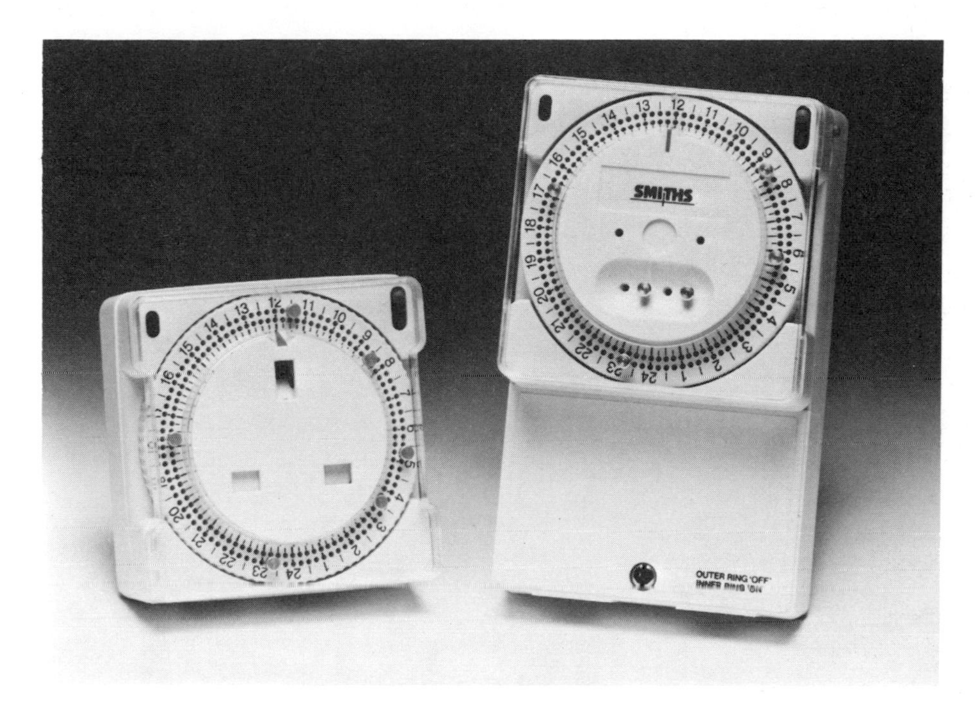

Figure 5.6 *Smith's timeswitch (plug-in type)*

illumination. Back entrances and exits are all too often overlooked as being unimportant, but it is precisely here that stolen goods are taken out of the premises. Stockrooms should also be well illuminated (*Figure 5.7*).

Whether external lighting should be directed towards or away from the building has its merits and demerits. The choice partly depends on architectural features. Away from the premises offers obvious advantages as, in order to gain entry, the intruder exposes himself to the full glare of the lamps. This is in contrast to the guards, who are positioned in the unlit area. On the other hand illuminating the exterior of the premises means that direct entry is clearly seen since any lit figure stands out. In both cases there is a sweep of light surrounding the building.

Running costs are reduced by the high-efficacy low-pressure sodium lamps, especially when wired on separate circuits, thereby allowing them to qualify for cheap night rates. The golden glow of SOX lamps is of a bright distinctive colour and is bound to act as a further deterrent by

Figure 5.7 *Stockroom security lighting*

warning that any non-uniformed person seen on the premises after working hours is probably an intruder.

Industrial premises

Properly organized security by illumination generally follows the means adopted for the commercial counterpart with special emphasis on safeguarding machinery, raw and finished materials, manufacturing processes etc. A beam of light directed to the entrances, exits and gateways offers a strong deterrent against taking out stolen goods. Consideration should also be given to inside and outside lighting of the perimeter as a means of providing 'belt and braces' security; it will certainly facilitate the spotting of unwanted intruders by staff guarding the premises.

It cannot be overemphasized that loss or damage to software can be extremely costly. Methods of production, organization, forward planning, salary differentials etc. all may be stored electronically. Security lighting plays an essential part in minimizing or completely eliminating damage or losses to such items which form an essential role in business enterprises.

Floodlighting

As an important security measure, large open spaces require a flood of light for the detection of unwanted persons. Such areas would include the larger plants, military establishments, aerodromes, goods yards, oil terminals, chemical works, car and lorry parks etc. Site conditions will dictate the type of floodlighting employed for each particular situation, which may be wall- or hollow-galvanized-pole- (to BS 729) mounted (*Figure 5.8*). The particular choice can be obtained from a wide variety of lamps and luminaires. Suitable lamps are fluorescent, tungsten halogen or high-pressure sodium (SON) rated at 50 W, 170 W or 120 W. The lamp as shown in the illustration when used with an appropriate reflector is suitable for medium- or long-range light projection.

An international symbol ⚠ indicates that the luminaire has an internal snap switch, so that it can be used whether or not they have an external starter fitted.

Reflector choice may be obtained for various types of light beams. The

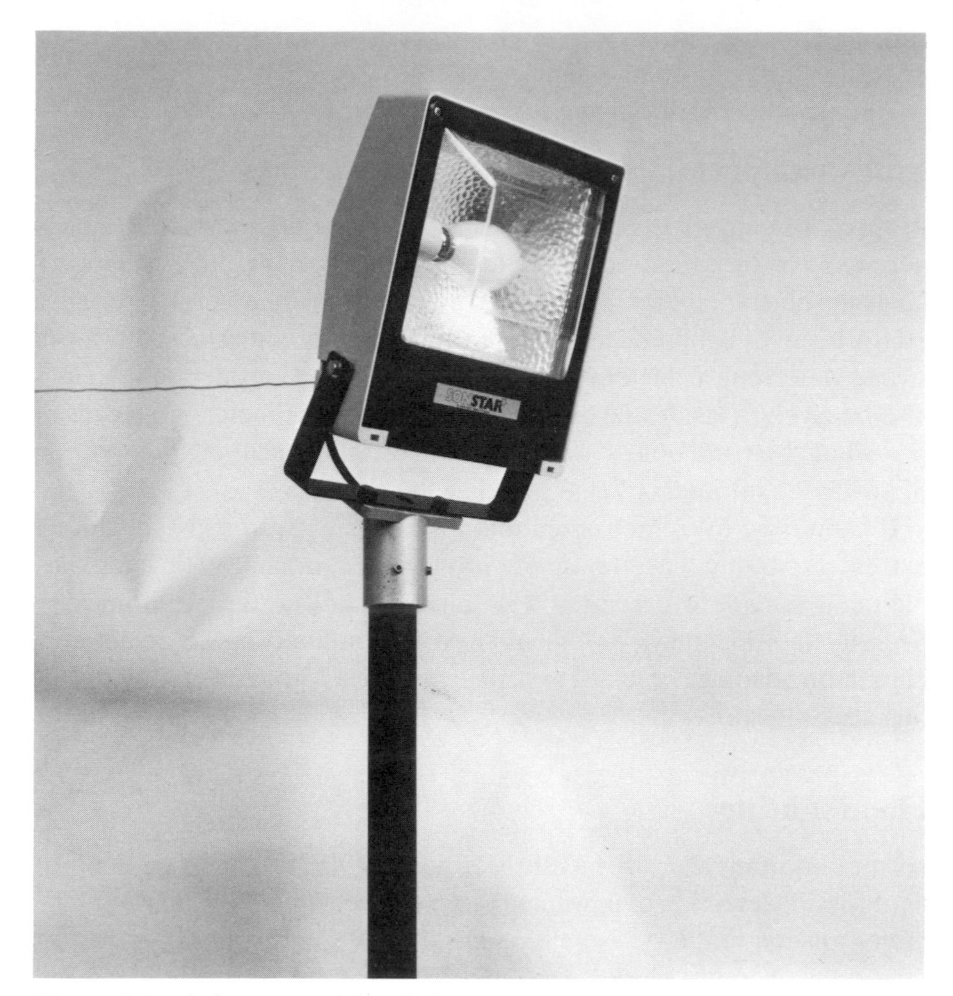

Figure 5.8 *Pole-mounted floodlight*

normal beam is parabolic, semicircular for symmetric beam or fan-shaped.

Reflectors are usually hammered or of dimpled finish to avoid dark and light lines. Wire guards give an added mechanical protection against stone-throwing or similar abuse. Most floodlight luminaires permit easy adjustment for beam angle, usually by wing nuts. By virtue of external positioning, the fittings are subject to the vagaries of our inclement weather, so that, in addition to a weatherproof finish, they should be manufactured to IP54 as being dust- and splash-proof.

Auxiliary areas which come under this category are, for example, local units for the examination of vehicles approaching the building doors and gates. Care must be taken that patrol guards are screened from the floodlit area.

Example 5.1

(a) In order to produce an illuminance of 10 lux for a floodlight area of 20 m × 60 m and assuming each lamp has an output of 5000 lumens, calculate the number of lamps and show typical layout.
(b) What is the power in watts per square metre under these conditions?

Solution

(a) From the equation as given in the Electricity Council's Guide, 'Essentials of Security Lighting',

$$\text{number of lamps} = \frac{\text{area}(\text{m}^2) \times \text{illuminance (lux)}}{\text{beam factor of luminaire} \times \text{lamp output (lm)} \times \text{maintenance factor}}$$

$$= \frac{20 \times 60 \times 10}{0.25 \times 5000 \times 0.8}$$

$$= 12.$$

(Note: A typical beam factor of 0.25 has been taken, but manufacturers must be approached in each case. It allows for spread of light, in contrast to concentration of light source. 0.8 is the usual figure for maintenance factor (now light loss factor), as an average value which may be varied in special cases.)

A suggested layout of the luminaires is shown in *Figure 5.9*.
(b) Selecting a GEC-OSRAM 70 W Sonstar lamp,

$$\text{W/m}^2 = \frac{\text{total power}}{\text{total area}}$$

$$= \frac{70 \times 12}{20 \times 60}$$

$$= 0.7.$$

Allowance must be made for ballast loss (about 15 W/lamp).

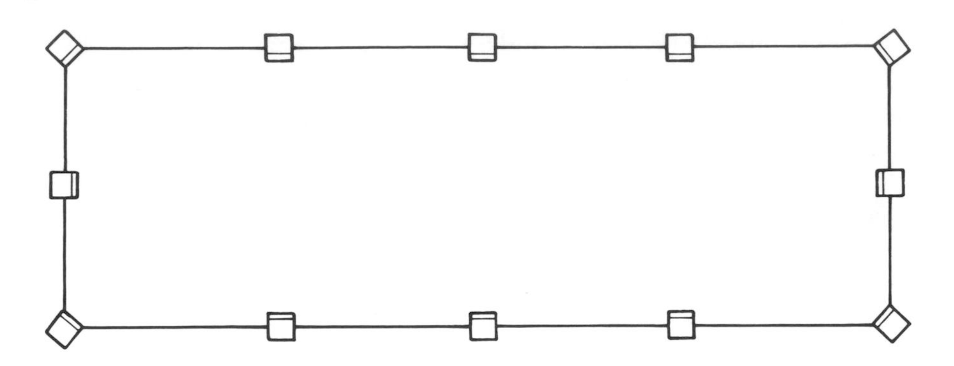

Figure 5.9 *Layout of floodlights (see example 5.1)*

Example 5.2

A rectangular area of 20 m × 12 m is to be illuminated by a luminaires pole mounted at each corner at a height of 10 m. If individual luminaires have an intensity of 24 000 cd directed to the centre of the area, what is the illuminance at the centre?

Referring to *Figure 5.10*, by Pythagoras' theorem:

each column to centre of area $(h) = (10^2 + 6^2)^{1/2} = 11.7$ m

$(d) = (10^2 + 11.7^2)^{1/2} = 15.4$

$$\cos\theta = \frac{h}{d}$$

$$= \frac{11.7}{15.4} = 0.76$$

Solution

Using the point source equation

$$\text{illuminance} = \frac{I\cos\theta}{d^2} \text{ where } I = \text{intensity in candelas}$$

$$= \frac{24\,000 \times 0.76}{15.4^2}$$

$$= 77 \text{ lux}$$

total illuminance from four luminaires

$$= 77 \times 4 \qquad = 308 \text{ lux}$$

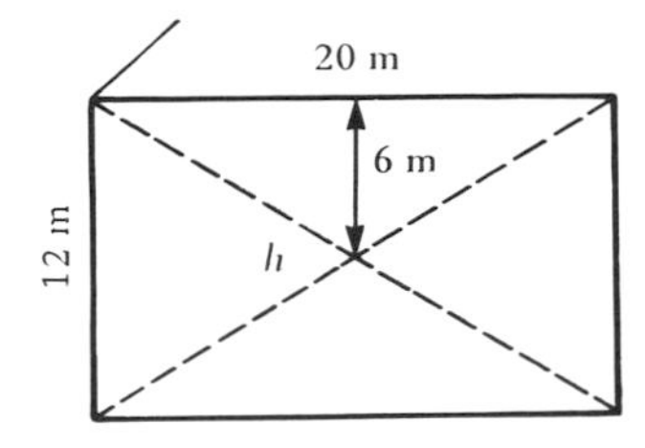

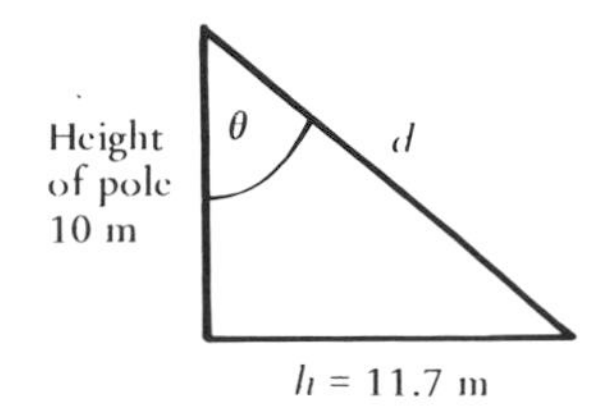

Figure 5.10 *Floodlighting point theorem (see example 5.2)*

Passive infra-red (PIR) detector

Here a PIR unit *Figure 5.11* is made to emit an invisible infra-red radiation, and persons within the infra-red area can be detected. The action is based on the principle that a person's heat within the area of the infra-red radiation operates the unit. The device is designed to switch on one or more lights immediately the area is entered. The unit includes a photoelectric cell to eliminate the possibility of inadvertent daytime operation.

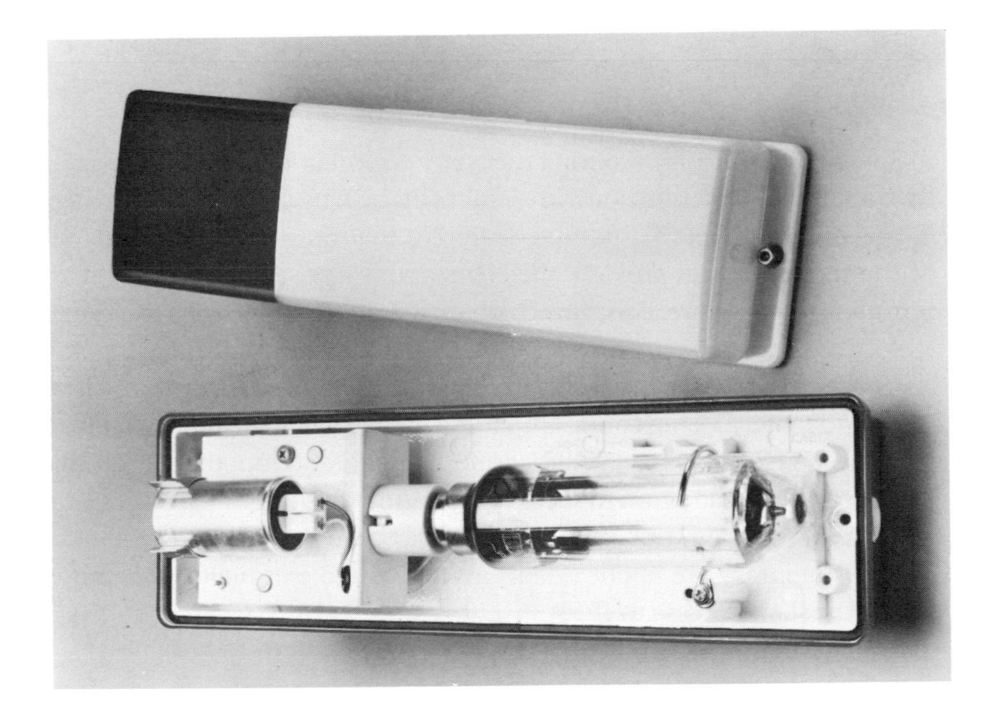

Figure 5.11 *Passive infra-red unit*

Figure 5.12 *PIM sensor unit*

Since no lighting is on during the 'safe' period of darkness, the use of PIR appears to offer obvious advantages by making considerable savings in fuel charges. Another positive feature is that the sudden switching on of the security lamps provides the element of surprise and, in contrast to normal lighting, glare assists in the detection of intruders.

As can be expected, many electrical manufacturers producing PIR units are exploiting the basic principle by supplying a multitude of variations and range of lamp controls up to 2 kW, and could include as many as six slave units. Lamps can act integrally or remotely from the sensor unit (*Figure 5.12*). In addition to security lamps the sensor unit may operate a built-in motor siren with a sweeping frequency to produce the maximum effect; soundings are about 120 dB. Where fitted externally, the unit is water-resistant if made to IP53.

The internal or external infra-red area range is 70° to 180°. *Figure 5.13* shows the total coverage when the PIR unit is positioned in a corner, although the actual radiation spread is in the nature of finger-like zones

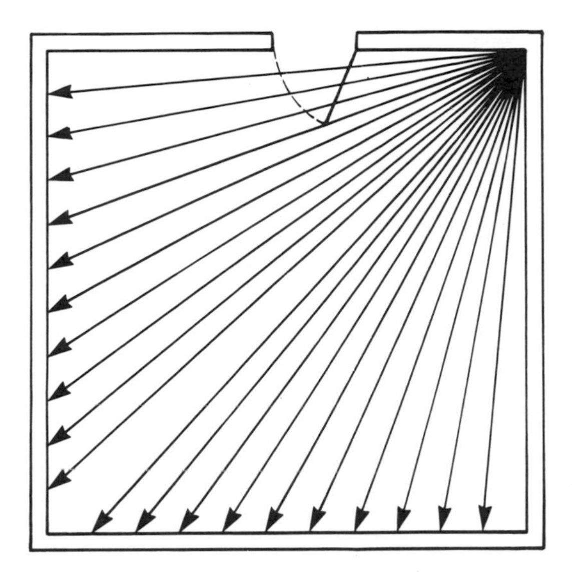

Figure 5.13 *Infra-red detector with 90° range*

rather than straight lines. Full coverage may be given internally if the sensor unit is fitted at the ceiling centre.

It is a common feature to fit PAR 38 spotlights or fluorescent compact lamps with the PIR. To minimize false alarms it must be arranged that stationary objects such as fixed heaters or moving curtains/doors do not affect the PIR unit. A person in the protected area sets off the alarm circuit for 1–4 minutes and a further time if remaining in the protected area.

When connected to a 12 V 1.2 Ah sealed rechargeable battery, operation will continue even if there is a mains failure as long as 100 h. On reconnection for 10 h full charge will be obtained.

As an indication of its value, the PIR unit has been accepted by a wide range of crime detection officials in conjunction with police, military and government authorities.

Closed-circuit television (CCTV)

As burglars' methods are becoming increasingly sophisticated, real safeguards can only be met by the implementing of more advanced methods. Hence, passive infra-red detectors can now be fitted in conjunction with closed-circuit television. This not only tends to frighten off the criminal but also, by videotapes, retains a permanent record of their physical features, including identification of intruder position and track movement.

The picture appears on the television screen immediately entry is made into the infra-red area. CCTV can give a 24 h protection, and the camera may be fitted with a zoom, focus or other ancillary lens. It may also integrate with automatic floodlighting.

Expensive items can be safeguarded by constant scanning with hidden cameras. Some types are particularly suitable where safeguarding against shop-lifting is deemed necessary. In one reported case, featured on BBC's television programme 'Crimewatch', a gang decided to 'risk it' and did not realize that their actions in stealing rings from a jeweller's shop window were being recorded on video. In addition, information reproduced from the tape recording appearing in the local newspapers made for positive identification of the defendants, who were later sentenced to serve a period of imprisonment. Publicity given to this and similar examples inevitably reduces the possibilities of robberies.

Many cameras have a built-in microphone as an added feature, so that when positioned at the entrance, the person at the reception desk can check credentials before permitting entry. For maximum security the siting of cameras in existing buildings requires very careful planning in order to cover selected points such as doors, entrances and exits. Some cameras can be set to cover a high number of zones. Also available are closed-circuit cameras which are linked with red flashing lights for the purpose of reinforcing the impression that full surveillance is obtained.

New structures give the opportunity of liaison with the architects with the object of making them lighting-security-minded, ensuring that cameras are positioned in the right places and the scanning avoids 'dead' areas.

Developments are continuing: large sites install a microprocessor to link scanning cameras; one advanced method uses integrated high technology for scanning wide areas by a satellite.

Where cameras should be placed is important. As noted in certain installations they should be hidden from view. In general the aim is to make them reasonably accessible yet at the same time free from possible tampering. On the other hand the cameras should not be positioned so remote as to make for difficulties in maintenance repair or replacement by future advanced types.

Certain acoustically immune sensors are able to guard against possible false alarms from external signals such as radio-frequency (RF) interference or transients from power lines. Planned CCTV acts as a powerful means of detection and saves the cost of employing one or more security guards.

Wiring

Maximum cable concealment is advisable by fitting mineral-insulated cables and heavy gauge steel conduits in walls and ceilings. The latter certainly minimize the possibility of supplies being cut off. All installations should comply with the latest edition and amendments of the IEE 'Regulations for Electrical Installations' and be of superior quality as, although the cables may be in use only for intermittent periods, they need to be in sound working condition when required. Thus the installation must be designed for high-integrity wiring.

Ingenuity in the fitting of ring or duplicate circuits can provide a further measure to minimize the risk of installation breakdowns, whether accidental or deliberate. Adjacent buildings sometimes afford entry through walls, ceilings and roofs. One company supplies a special guardwire to prevent this type of entry. During building construction sensor cables are embedded in walls, ceiling or floors, the latter to prevent tunnelling from below. In existing structures these sensor cables are carefully run on walls and attached to ceiling and roof load-bearing joists.

As a further safety measure, the entire earth path may also be continuously monitored. In one system this is achieved by circulating an earth current in the order of 1 A at 12 V. The monitoring current circulates through a loop circuit consisting of a relay, the main circuit protective conductors and the return path.

In the event of damage to the circuit-protective conductors or any part of the earth system resulting in a loss of continuity in the loop circuit, the relay becomes de-energized and operates the trip mechanism of a circuit-breaker. The circuit is thereby disconnected and cannot be reconnected until the earth fault is cleared. Essential to this method is that two independent earth terminals are provided to ensure that lamp units form part of the monitored loop. A further development is the combined earth-leakage monitoring unit for transportable and portable equipment.

Electric shock from direct contact is obtained by simultaneously touching phase and neutral conductors or phase and earthed metalwork. Protection is provided generally by insulation, barriers or obstacles, or placing the wiring out of reach. However, in the case of exposed conductive parts (metal casing of switchgear etc.), which themselves are not intended to carry current but can become live under faulty conditions, protection is obtained by earthed equipotential bonding and automatic disconnection of supply, often referred to as 'EEBAD'. Protective devices consist of fuses (rewirable or of high breaking capacity), miniature circuit-breakers (MCB), moulded-case circuit-breakers (MCCB) or circuit-breakers. High breaking capacity is rated as an essential requirement of these safety devices since short-circuit currents can be of extremely high value. A short-circuit may be defined as a direct connection between phase and neutral conductors. On a 240 V single-phase circuit if, on a short-circuit, the resistance between conductors is 0.01 Ω, then the prospective short-circuit current could rise to 240 V/0.01 Ω = 24 000 A!

Surges from inductive circuits, without power-factor correction, have been known to blow holes in the sheath of mineral-insulated cables. A solution is to fit surge diverters to inductive loads in order to suppress possible damage caused by these high voltages.

Shock tactics

Apart from earthing and definite insulation of live parts, any additional measure may be adopted to ensure protection against electric shock is by the use of safety extra-low voltage (SELV), where the nominal voltage does not exceed 50 V AC or 120 V DC indirect contact and supplied by one of the following:

1 secondary winding of a Class II isolating transformer conforming to BS 3535. On this secondary side there must be no connection to earth
2 a source of current providing safety equivalent to a safety isolating transformer, e.g. a motor-driven generator giving equivalent isolation
3 a battery or other source independent of a higher-voltage circuit
4 an electronic device so designed that even under internal fault conditions extra-low voltage must not be exceeded at the output.

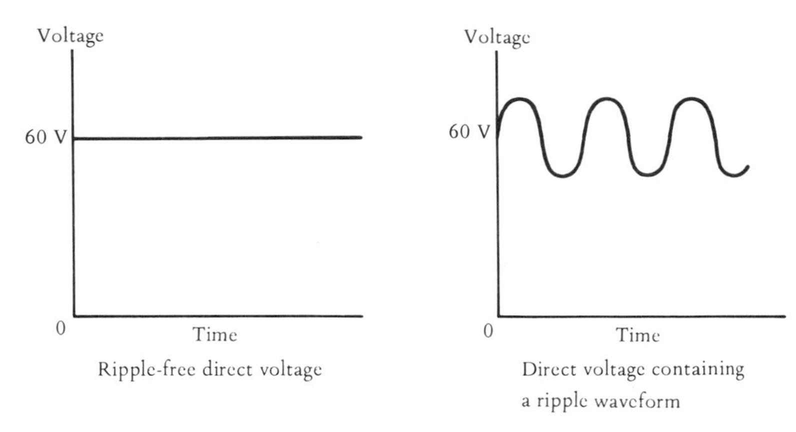

Figure 5.14 *Graphical representation of direct voltages*

Class II denotes a primary and secondary layer of insulation (i.e. double insulation) and applies to a growing number of items of electrical equipment, which in the interest of safety, does not require to be earthed and is applicable to many types of luminaire.

SELV cables must not be in contact with normal mains cables. Thus, when both types are following the same route this arrangement would normally be achieved by PVC trunking with separate partitions.

From the above it follows that plugs fed by SELV are to be non-reversible 2 pin, or alternatively with an unearthed third pin. These plugs must also be so designed that they are incapable of entering the socket outlets supplied by other voltages.

In situations where there is a possibility of direct contact the SELV voltage must not exceed 25 V 50 Hz AC or 60 V ripple-free DC (*Figure 5.14*), and in addition conform to the following two conditions:

1 barriers or enclosures affording protection to at least IP2X (BS 5490)
2 circuits must be subject to a test voltage of 500 V for 1 minute without any breakdown of insulation.

Voltage drop

The term refers to the fall in voltage along cables supplying an electrical load (here lamps) and applies to emergency lighting as well as security lighting. It is essentially, by Ohm's law, an *IR* drop, where *I* is the current in amperes taken by the lamps and *R* is the resistance in ohms of the cable

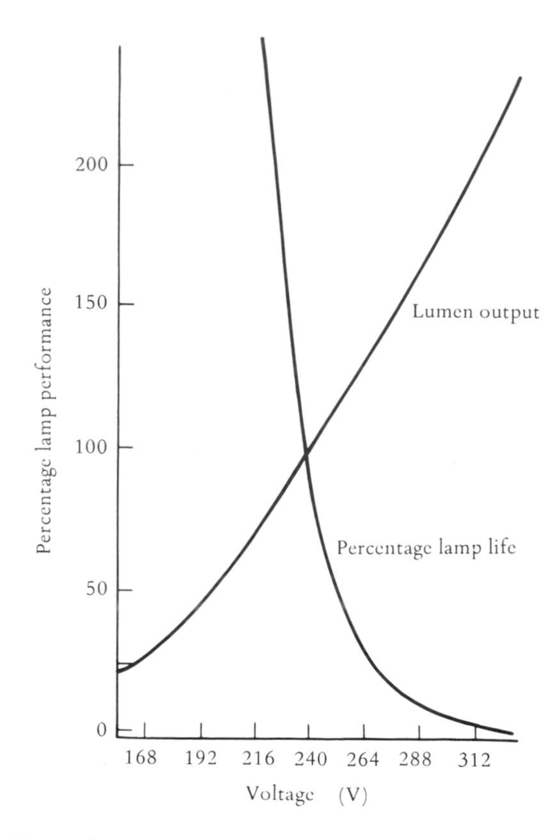

Figure 5.15 *Typical GLS performance curves*

resistance (live and neutral for single-phase loads) which acts, in effect, as a series resistor. Clearly, if the potential difference (voltage) at the lamp terminals is lower than the lamp's rated voltage then there will be a drop in light output. A typical performance graph for a GLS lamp is shown in *Figure 5.15*.

All conductors offer some resistance to the flow of current. The longer the conductor the greater is its resistance, so that resistance is proportional to conductor length. On the other hand, a thicker wire permits an easier flow of current, so that resistance can also be understood is being inversely proportional to its cross-sectional area.

As a guide, the IEE Wiring Regulations advised that the voltage drop should not exceed 2.5%. On a standard 240 V circuit this means that the voltage at lamp terminals should not fall below $100 - 2.5/100 \times 240$ V, i.e. 234 V; in other words, the voltage drop under these circumstances should not exceed 6 V.

In practice, voltage drop problems are solved by referring to the IEE Wiring Regulations tables, which give current-carrying capacity and voltage drop in terms of millivolt per ampere per metre length of run (mV/I/m). It is important to note that the tables take into account, by the use of correction factors, the actual conditions under which the cables are run.

Example 5.3

A 30 m run of single-core PVC non-armoured copper cable is housed in PVC conduit with one side of the conduit positioned against thermal lagging in an ambient (surrounding) temperature of 35°C. What is the minimum size of these cables to supply a 5 kW bank of floodlights? Protection is to be provided by an appropriate-sized miniature circuit-breaker.

Solution

$$\text{power} = \text{voltage} \times \text{current}$$

$$\therefore \text{current} = \frac{\text{power}}{\text{voltage}}$$

$$= \frac{5000}{240} = 20.83 \text{ A}$$

Nearest-size MCB = 25 A

General equation for obtaining current rating

$$= \frac{\text{fuse or MCB rating}}{C_1 \times C_2 \times C_3 \times C_4}$$

The correction factors are

C_1 = grouping
C_2 = ambient temperature
C_3 = thermal insulation
C_4 = type of protection

Here there is no grouping,

$$\begin{aligned} \therefore C_1 &= 1 \\ C_2 &= 0.94 \text{ (from IEE Appendix 9, Table 9D1)} \\ C_3 &= 0.75 \text{ (0.5 if totally immersed in lagging)} \\ C_4 &= 1 \text{(0.725 for rewirable fuse)} \end{aligned}$$

$$\text{Required cable rating} = \frac{25}{0.94 \times 0.75} = 35.5 \text{ A}$$

From IEE Table 9D1, col. 2, 6 mm^2 cable carries 41 A
Testing for voltage drop,

$$\frac{\text{mV/A/m} \times I \times 1}{1000} = \frac{7.1 \times 20.83 \times 30}{1000} \text{ (7.1 from col. 3)}$$

$$= 4.44 \text{ V}$$

This is below 6 V $\therefore$ selected cable size 6 mm^2.

Example 5.4

A security building is to be constructed at the entrance to a factory. This new building is to be provided with a 240 V single-phase supply and is to be situated 20 m from the main switchroom. A 30 m twin PVC-insulated armoured underground cable (copper conductors) supplies the new building, which allows 5 m at each end for runs within the main switchroom and security buildings.

The connected load in the security building comprises
one 3 kW convector heater
two 1 kW radiators
two 1.5 kW water heaters
one 6 kW cooker
six 13 A socket outlets (ring circuit)
a 2 kW lighting load
Diversity may be applied (business premises).
(a) Establish the minimum current rating of the switch fuse in the switch room at the origin of the main cable.

(b) Describe the precautions when the underground enters the building.
(c) Determine
(i) the minimum current rating of PVC-insulated twin-armoured (copper conductor) underground cable, assuming an ambient temperature of 25°C and protection to BS 88, Part 2
(ii) the minimum size of cable, assuming voltage drop is limited to 2 V
(iii) the actual voltage drop in the cable
(City and Guilds 236 course C. The accuracy of the question and answer are the sole responsibility of present author.)

Solution

(a) Total load applying diversity as necessary from IEE Wiring Regulations Appendix 4, Table 4B:

$$\text{one 3 kW convector heater} = \frac{3000\text{ W}}{240\text{ V}} = 12.5\text{ A}$$

$$\text{two 1 kW radiators} = \frac{2000}{240} \times \frac{75}{100} = 6.25\text{ A}$$

$$\text{two 1.5 kW water heaters} = \frac{3000}{240} = 12.5\text{ A}$$

$$\text{one 6 kW cooker} = \frac{6000}{240} = 25\text{ A}$$

$$\text{ring circuit} = 30\text{ A}$$

$$\text{2 kW lighting load} = \frac{2000}{240} \times \frac{90}{100} = 7.5\text{ A}$$

$$\text{total current} = 93.75\text{ A}$$

$$\therefore \text{minimum current rating of switch-fuse} = 100\text{ A.}$$

(b) When entering the building, internal radius of cable bend must not be less than six times the overall cable diameter. Fixings must ensure that the cable does not undergo any mechanical stress and the entry surrounding hole must be made good with a fire-resistant material.

(c) (i) Minimum cable current rating $= \dfrac{\text{fuse or MCB rating}}{\text{correction factor(s)}}$

$$= \frac{100}{1.02}$$

$$= 97.1 \text{ A.}$$

(ii) 1.02 is the correction factor extracted from Table 9C1 and relates to an ambient temperature of 25°C.

$$\text{Voltage drop} = \frac{\text{mV/A/m} \times I \times l}{1000}$$

$$= \frac{1.75 \times 93.75 \times 30}{1000}$$

$$= 4.92 \text{ V too high}$$

(where I = actual current flowing, l = length of run in metres, and 1000 as a denominator converts millivolts to volts). 1.75 is the mV/A/m in Table 9D4 and relates to 25 mm^2 conductor which carries 118 A. The cable is assumed to be purely resistive.

(iii) Try 70 mm^2, voltage drop $= \dfrac{0.63 \times 93.75 \times 30}{1000}$

$$= 1.77 \text{ V}$$

∴ minimum size cable is 70 mm^2 and actual voltage drop is 1.77 V.
Comments: The maximum volt drop of 2 V for the feeder cable allows a PD of 4 V for the distributor cables.

Voltage drop on extra-low (ELV) circuits

Certain tungsten halogen and some discharge lamps may have advantages by being run off ELV supplies. Under these conditions voltage drop becomes of critical importance in its effect on lumen output. A 2.5% voltage drop for a 50 W 12 V circuit is as little as 0.3 V.

Where these lamps are run from individual transformers and with the use of 1 mm^2 conductor the maximum cable length is 3.5 m. This extends to 5.1 m and 8.4 m for 1.5 mm^2 and 2.5 mm^2 conductor cross-sectional areas, respectively. Should the lamps be directly fed off individual transformers, this problem does not arise.

ELV cables must not come in contact with 240 V cables unless the former insulation is equal to mains voltage insulation.

The transformer should be made to double-insulated class II, BS 3535. In this connection several practical points must be considered:

1 terminals on the ELV side must be large enough to accommodate possible large size of cables
2 accessibility
3 adequate support
4 as there is bound to be a certain amount of heat, facilities for adequate ventilation must also be provided
5 the SELV system ensures additional safety so that, if adopted, the transformer secondary and the whole of the ELV portion, including lamps, must not be earthed.

6

Maintenance and economics

General features

Maintenance is normally a management responsibility, although it must be understood that according to the Health and Safety at Work Act 1974, workers also have responsibility and are required to report faulty equipment which may cause danger.

All forms of electrical equipment are subject, in time, to deterioration. Artificial lighting, in both emergency and security forms, is adversely affected over a period of time and a drop in illuminance occurs unless corrective action by proper maintenance is taken. This may be thought to be expensive.

However, a sound and reliable lighting installation may even pay dividends as indicated in *Table 6.1*:

Replacement of worn lamps is bound to increase light output. Safety is also improved by the removal of faulty wiring and fitments.

Table 6.1 *Positive features offered by maintenance*

—out-of-date luminaires and lamps replaced by modern types
—improved illumination tends to increase output of firm's products
—better working conditions
—improves safety and security
—makes emergency paths more definite
—gives opportunity to replace inefficient lamp units with ones requiring lower energy costs

Treatment for electric shock

Staff should be trained in the treatment for electric shock. The St John's Ambulance poster, 'Emergency resuscitation', advises the following order of action:

1 switch off if possible
2 secure release of victim
3 start resuscitating if not breathing
4 commence external cardiac massage if heart has stopped beating
5 send for doctor and ambulance.

The current should be switched off immediately, but if this is not possible, do not waste time searching for the isolator. Speed is essential as a severe shock affects the controlling nerves for breathing and heart action.

When securing release of the casualty it is essential to safeguard the person who is carrying out the artificial respiration from receiving a similar shock. One should stand on a non-conducting material (e.g. rubber mat, dry wood or dry newspaper). A length of dry rope may be employed to move the victim clear of contact with the live conductor.

Should no signs of breathing be indicated, remove impediments, such as a tight collar, and start artificial resuscitation straight away. Mouth-to-mouth resuscitation, sometimes called the 'kiss of life', is now considered a preferred method, and is described as follows. Places the casualty on his back and tilt the head fully backwards. The casualty should have the maximum air passage by fully opening the mouth. Seal the nose by pressing the nostrils together and apply the mouth-to-mouth procedure. Take a deep breath, place the mouth over the victim's mouth and breathe into the lungs. The mouth is then removed and with a deep breath the operation is repeated for about 6 to 10 s until there are signs of definite recovery or the doctor signifies that the operation should cease.

Maintenance should be coupled with proper supervision at set periods but must not militate against the replacement or rectification of faulty equipment as required. This may range from the replacement of a malfunctioning local switch to complete rewiring of the installation. The latter is essential if the wiring is more than 30 yr old and has been carried out with vulcanized india-rubber (VIR) cables.

The importance of maintenance cannot be overemphasized. It will not only rectify faults as they occur, but perhaps, even more importantly, should aim to (a) maintain the installation in a satisfactory condition and (b) remove causes of potential trouble before they actually occur. If these actions are taken then the frustration due to losses by bad working or even non-working of lamps will be avoided.

Large firms will employ one or more maintenance engineers charged with the responsibility of understanding the function of each item of equipment and therefore may need, from time to time, to attend courses and liaise with outside specialists. Certainly maintenance staff need to keep up-to-date with modern energy and security lighting techniques.

The IEE Wiring Regulations have references to the fundamental safety aspects of maintenance (Reg. 13-2 and Reg. 341-1). Good workmanship and proper materials—wherever possible complying with British Standards—are a prerequisite to the type of installation which would permit proper maintenance.

Efforts to obtain security are sometimes frustrated by break-ins resulting from inside information or assistance. There is no perfect answer to such criminal acts although they can be minimized by thoroughly examining and checking references to ensure the integrity of personnel responsible for undertaking the work.

For the larger industrial sites a decision has to be formulated as to whether it is preferable to employ the firm's own maintenance engineers or to give the work to outside contractors. The latter has the disadvantage that it includes an unknown factor and therefore is a security risk.

Ladders

With luminaires out of reach the use of a ladder may be necessary. Some hints on safe use are:

1 before using the ladder, examine for defects which may render it unsafe
2 the ladder is to be set on firm even ground at 75°, i.e. 4 up and 1 out
3 if possible secure at the top with ties at the stiles (sides)
4 it is recommended to have one man at the bottom with one foot on the ground and the other on the first rung while holding the stiles
5 care is to be taken against swaying when climbing the ladder.

To avoid a dangerous jerk, step-ladders must be fully extended before use.

Inspection and testing

Physical inspection must be supplemented by use of a lightmeter, especially for emergency lighting, as separate from general lighting, at various relevant positions to ascertain whether the illuminance is sufficient. For accuracy the lightmeter (photometer) should be checked annually.

In addition to the verification of the light level a number of electrical installation checks should be made during the design stage such as selection of conductors for current-carrying capacity and voltage drop, especially for lamps operated by extra-low voltage. Others, e.g. polarity, should be examined during actual testing.

Examination of weak links in installations

1 flexible cords where connection to
  (a) plugs and apparatus
  (b) light pendants
2 batten holders with high-powered filament lamps vertically below them
3 physical inspection is essential to ensure soundness of luminaires, i.e. no cracks or rust. All items should be firmly fixed and there are to be no avoidable bare conductors at terminal connections
4 comprehensive labelling is important. In addition, attention should be drawn to the need for appropriate diagrams and instructions.

Recommended testing sequence:

1 continuity and actual ohmic value of protective conductors
2 earth electrode resistance
3 insulation resistance
4 barriers and enclosures
5 polarity
6 earth-fault loop impedance
7 operation of residual-current circuit-breakers
8 emergency lighting follows clear escape routes
9 a number of night visits to observe consistent operation of security lighting.

Here, inspection and testing are stated in general terms. Each unit requires its own individual attention and procedure.

When carrying out electrical tests it is important to appreciate that a contactor, by itself, is not sufficient to serve as a safe means of isolation.

Lamps on a maintained system clearly show when they are in operation, in contrast to the non-maintained single-point conversion units which are only illuminated by a failure of supply. Thus these latter types require periodic manual checking. Alternatively, certain manufacturers market a means of automatic testing rendering obvious advantages.

Residual-current circuit-breakers (RCD) also require regular checking by specialized equipment to indicate that they are operating correctly.

Planned maintenance

One of the purposes of correct planning is to facilitate that the maintenance of these specialist items of equipment is in sound working order. Many variations will have to be taken into account, such as different methods to be adopted for emergency and security lighting, while noting that there are certain common lamp features.

A basic approach depends on whether there is to be a completely new installation or if simply modification is required to an existing scheme—or even that there is an absence of a scheme! The possession of any past records will be of vital importance in determining what changes are necessary and will serve as a check on replacements. Installations need the co-operation of architects for optimum results in planning escape routes for emergency lighting and positioning of security lamps.

To comply with local regulations, liaison with appropriate authorities is a must. Incidentally, a great deal can be learned from installations in other buildings.

In many cases maintenance is to be regarded as of equal importance to the original work. Lighting controls must be accessible and not tucked behind belts, chains or machines.

It is essential that luminaires conforming to IP 5X and IP 6X, which are dustproof and dust-light respectively, are to be installed in dirty atmospheres. These types are self-ventilating by air currents to achieve cleaning action and are an improvement on the 'bucket and water' (soapy solution) application. If, as an alternative, jet spraying is employed then great care must be taken to guard against water and moisture penetration. It cannot be overemphasized that dust and dirt reduce light output.

In planning, a crucial point to observe is lamp replacement. At the end of their useful life the filaments of both general service (GLS) and tungsten halogen (TH) tend to snap without warning, thereby plunging the area into darkness. On the other hand, discharge types flicker or show a gradual drop in light output. Lamp manufacturers cite hours of useful life of their products, but will be shortened if there is too frequent switching ON and OFF.

Figure 6.1, reproduced from the CIBSE Code for Interior Lighting, produced by permission of the Chartered Institution of Building Services Engineers, indicates changes in illumination by different lamp replacement and cleaning schedules for tubular fluorescent lamps. The general pattern will change for different lamps, but the underlying principle will be unaltered.

Figure 6.1(a) depicts the normal graph for fluorescent lamps left to run out completely. Reduced losses (*Figure 6.1(b)*) occur by cleaning every 12 months and compare with further reduction by 6 monthly periods (*Figure 6.1(c)*). The final diagram (*Figure 6.1(d)*) indicates the beneficial effects obtained by 6-monthly lamp cleaning and 2-yearly replacement of lamps.

Cleaning and lamp replacement is a managerial decision, and again highlights the importance of keeping proper records. Certainly the position should be reviewed at not later than 12-monthly intervals. While a lamp may fail after a short period, group replacement is generally more economical than one-by-one replacement.

Bare lamps or luminaire glass covers only require cleaning with a damp cloth or sponge, but the cleaning method may be supplemented by a detergent which in turn requires rinsing. Finally the surface should be polished with a clean soft cloth.

The above method may also be used for plastics. However, owing to the accumulation of dust by static charges, use of an antistatic fluid produces improved results.

Battery maintenance

To maintain a battery in good condition hints to note are:
1 newly purchased batteries should be carefully unpacked and detailed manufacturer's instructions followed

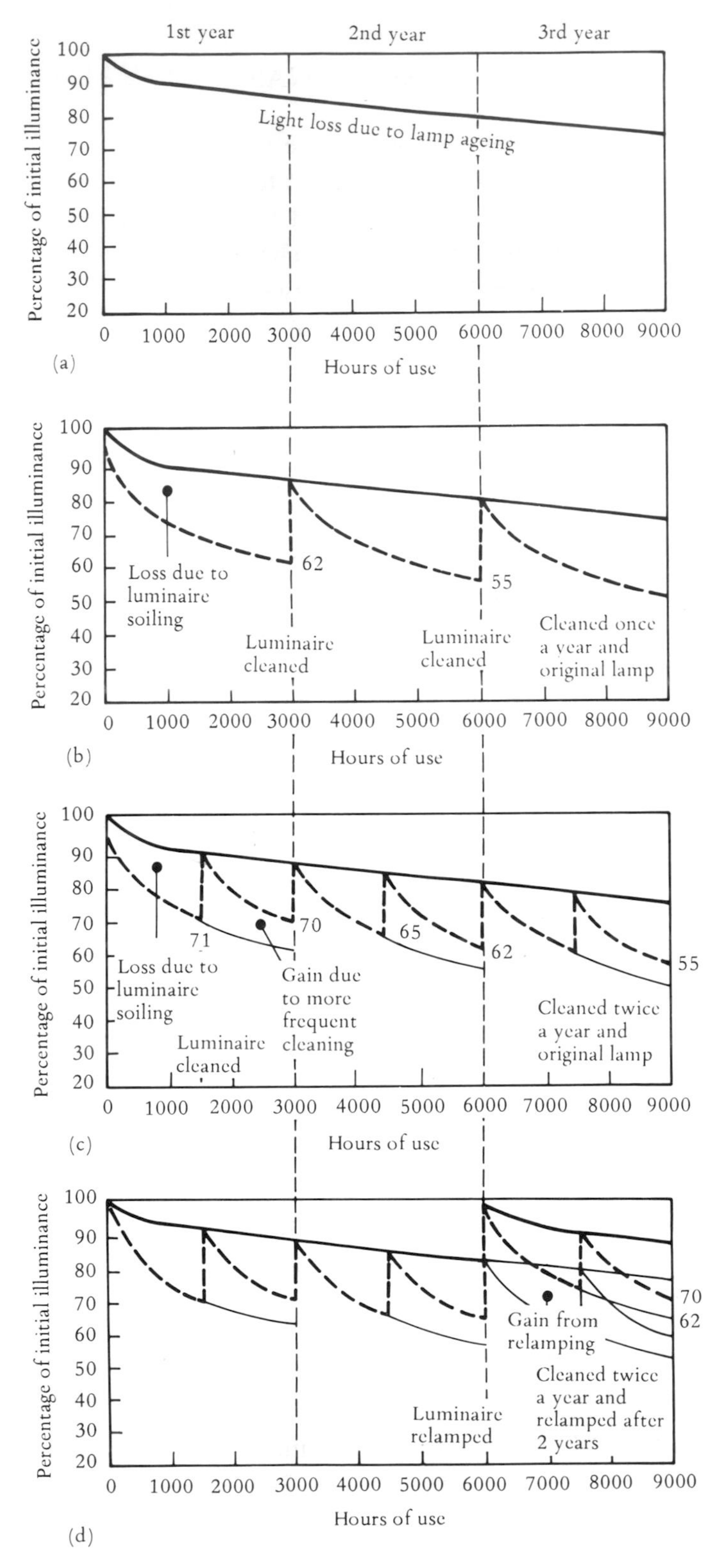

Figure 6.1 *Changes in illuminance by different maintenance schedules*

2 lead-acid types must be filled with pure sulphuric acid to BS 3031. Allow for evaporation—pour distilled water into the acid and not sulphuric acid into distilled water. Liquid level must cover the entire plates. The cell voltage should be 2.7 V on charge. If it falls to below 2.3–1.9 V there is likely to be a short-circuit between plates, accompanied by a rapid drop in the density of the electrolyte
3 cleanliness is essential. Acid impurities are bound to reduce the working life. Care must be taken to clear vent plugs, otherwise internal pressure may be built up
4 proximity of naked lights could easily ignite battery gas, causing a violent explosion
5 the battery should be kept in dry, cool premises and must be accessible for testing and topping up
6 regular checks must be made to ensure operation of emergency lighting and generator starting
7 cells should not be allowed to stand in an uncharged state
8 lead-acid sealed and nickel-cadmium batteries require the minimum of maintenance, but outputs will still need to be regularly tested.

Generator maintenance

In addition to the regular electrical tests (monthly when not run continuously), visual inspection may reveal rusting, cracked or broken parts, loose connections and illumination of indicator or pilot lamps.

Among periodic mechanical checks are:
1 engine oil pressure
2 high-water shut-down relays (where fitted)
3 any loose fit of windings in slots caused by general coil vibration
4 loose fit of wedges in stator slots
5 overheating of windings, producing swelling and deformation
6 abrasion by incorrect fit and/or carbon particles
7 misalignment of bearings.

Costing

It does not follow that an initital saving in emergency and security lighting equipment brings a lower total costing. Persons responsible for the

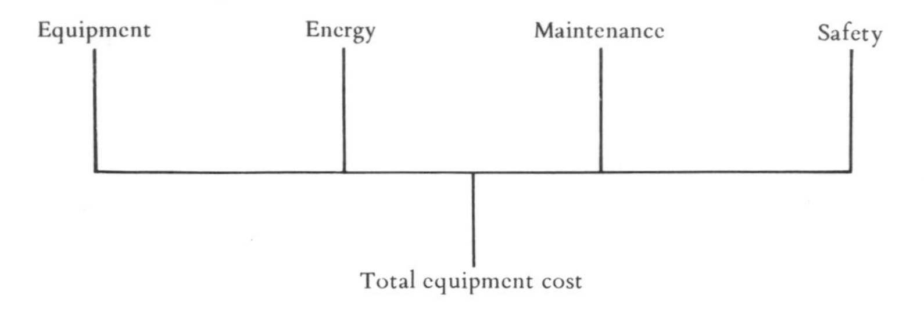

Figure 6.2 *Costing*

buying must take into account four basic concepts (*Figure 6.2*) before purchasing the equipment. For example, compact fluorescent lamps may bring an 80% reduction in maintenance charges as compared with GLS lamps and could effect a cost recovery in two years. Electric lamp technology is in continuous development, and in this competitive world forward-looking management will seek out appropriate units effecting an overall saving in electrical energy.

Fundamental energy charges are on the 2-part tariff consisting of (a) a fixed charge based on a number of variables, e.g. total floor area and (b) a running energy cost. It is in this latter aspect that considerable savings may be made without lowering the standard of the installation. If electricity boards cite a 'power factor clause', then fluorescent tubes operated by the electronic ballast show a reduction in costs as compared to the 'conventional' copper/ballast since the former gives a power factor of near unity.

Since peak loads for short periods increase generator (and its ancillary) costs, care should be taken when considering the 'maximum demand' tariff for floodlighting.

With an extensive site area, there may be group maintenance programmes requiring the use of one or more computers which can itemize details of this specialized lighting, enabling a reduction in labour. Breakdowns can thus be minimized since incipient faults will be revealed at an early stage.

References

Lighting Industry Federation. *Lamp Guide*
Lighting Industry Federation. *Hazardous Lighting*
BS 161 Tungsten filament lamps for general service
Code of Practice 1013, Earthing
BS 185 Tubular fluorescent lamps for general service
BS 229 Flameproof enclosure of electrical apparatus
BS 667 Portable photoelectric photometers
BS 3677 High pressure mercury vapour lamps
BS 3767 Low pressure sodium vapour lamps
BS 4533 Luminaires. Includes IP ingress protection system
BS 5266: Part 1: 1988 Code of practice for the emergency lighting of premises
BS 5489 Code of practice for roadlighting
IEC Publication 662. High pressure sodium vapour lamps
Technical Memorandum TM5. The calculation and use of utilisation factors
Technical Memorandum TM10. Evaluation of discomfort glare
Electricity Council. *The Essentials of Security Lighting*
Health and Safety Executive. *Emergency Private Generation: Electrical Safety*
The Institution of Electrical Engineers. *Regulations for Electrical Installations*

Discharge lamps symbols:
M—mercury

B—quartz arc tube
F—fluorescent coating
W—woods black glass
T—tungsten filament ballast
R—internal reflector
U—universal operating position
V—vertical cap up
D—vertical cap down
I—iodide

Index